# ARM
# Architecture
# Reference
# Manual

Document Number: ARM DDI 0100B
Copyright Advanced RISC Machines Ltd (ARM) 1996

**ENGLAND**
Advanced RISC Machines Limited
Fulbourn Road
Cherry Hinton
Cambridge CB1 4JN
UK
Telephone:   +44 1223 400400
Facsimile:   +44 1223 400410
Email:       info@armltd.co.uk

**JAPAN**
Advanced RISC Machines K.K.
KSP West Bldg, 3F 300D, 3-2-1 Sakado
Takatsu-ku, Kawasaki-shi
Kanagawa
213 Japan
Telephone:   +81 44 850 1301
Facsimile:   +81 44 850 1308
Email:       info@armltd.co.uk

**GERMANY**
Advanced RISC Machines Limited
Otto-Hahn Str. 13b
85521 Ottobrunn-Riemerling
Munich
Germany
Telephone:   +49 89 608 75545
Facsimile:   +49 89 608 75599
Email:       info@armltd.co.uk

**USA**
ARM USA Incorporated
Suite 5
985 University Avenue
Los Gatos
CA 95030 USA
Telephone:   +1 408 399 5199
Facsimile:   +1 408 399 8854
Email:       info@arm.com

World Wide Web address: http://www.arm.com

**PRENTICE HALL**
London  New York  Toronto  Sydney  Tokyo  Singapore
Madrid  Mexico City  Munich  Paris

## Proprietary Notice

ARM and the ARM Powered logo are trademarks of Advanced RISC Machines Ltd.

Neither the whole nor any part of the information contained in, or the product described in, this document may be adapted or reproduced in any material form except with the prior written permission of the copyright holder.
The product described in this document is subject to continuous development and improvement.
All particulars of the product and its use contained in this document are given by ARM in good faith.
However, all warranties implied or expressed, including but not limited to implied warranties or merchantability, or fitness for purpose, are excluded.

This document is intended only to assist the reader in the use of the product. ARM Ltd shall not be liable for any loss or damage arising from the use of any information in this document, or any error or omission in such information, or any incorrect use of the product.

## Credits

Thanks are due to:

Allen Baum, Richard Earnshaw, Jim Eno, Pete Harrod, Barbara Henighan, Greg Hoepner, Lance Howarth, Guy Larri, Neil Robinson, David Seal, Lee Smith, Ray Stephany, John Webster, Sophie Wilson, Rich Witek

## Change Log

| Issue | Date | By | Change |
|---|---|---|---|
| Draft | 28 Jul 95 | DJ/BJH | Created |
| A | 7 Feb 1996 | BJH | Review comments added. |
| B | 26 July 1996 | BJH | Updated release. Index added. |

Printed in Great Britain by T J International Ltd

ISBN 0-13-736299-4

**ARM Architecture Reference Manual**
ARM DDI 0100B

**The ARM architecture is the basis of the world's most widely available 32-bit microprocessor.**

ARM Powered microprocessors are being routinely designed into a wider range of products than any other 32-bit processor. This diversity of applicability is made possible by the ARM architecture, resulting in optimal system solutions at the crossroads of high performance, low power consumption, and low cost.

In November 1990, ARM was formed to develop and promote the ARM architecture. With initial investment from:

- Apple Computer, the world's third largest manufacturer of personal computers
- Acorn, the United Kindom's leading supplier of Information Technology for education
- VLSI Technology, the world leading ASIC supplier

ARM Ltd. began work to establish the ARM architecture as the 32-bit standard microprocessor architecture.

Initially, ARM devices were made by VLSI Technology, but ARM's business model is to license microprocessor designs to a broad range of semiconductor manufacturers, allowing focus on a variety of end-user applications. To date, this licensing strategy has resulted in twelve companies manufacturing ARM-based designs:

| | | |
|---|---|---|
| VLSI Technology | GEC Plessey Semiconductors (GPS) | Sharp Corporation |
| Texas Instruments (TI) | Cirrus Logic | Asahi Kasai Microsystems (AKM) |
| Samsung Corporation | Digital Equipment Corporation | European Silicon Structures (ES2) |
| NEC Corporation | Symbios Logic | Lucky Goldstar Coporation |

Through these twelve semiconductor partners, the ARM processor is currently being used in a wider range of applications than any other 32-bit microprocessor. ARM chooses new partners carefully; each new partner is judged on their ability to extend the applicability of ARM processors into new market areas, broadening the total ARM product range by adding their unique expertise.

Customers using ARM processors benefit not only from the ARM architecture, but the ability to select the most appropriate silicon manufacturer. Furthermore, the worldwide awareness of the ARM processor has attracted third-party developers to the ARM architecture, yielding a huge variety of ARM support products:

ISI, Microtec, Accelerated Technology, Cygnus, Eonic Systems and Perihelion all offer Operating Systems for the ARM.

Yokogawa Digital and Lauterbach provide In-Circuit Emulators (ICE).

Hewlett Packard Logic analysers support ARM processors.

In ARM Ltd's five-year history, it has delivered over 30 unique microprocessor and support-chip designs. ARM microprocessors are routinely being designed into hundreds of products including:

| | |
|---|---|
| cellular telephones | interactive game consoles |
| organisers | disk drives |
| modems | high performance workstations |
| graphics accelerators | car navigation systems |
| video phones | digital broadcast set-top decoders |
| cameras | smart cards |
| telephone switchboards | laser printers |

# Foreword

This product range includes microprocessors designed as macrocell cores, with around 4mm$^2$ of silicon, drawing less than fifty milliWatts of power when executing over 30 MIPS of sustained performance. For very high performance applications, ARM implementations deliver over 200 MIPS sustained performance (more than most high-performance computer workstations), while still only consuming one half of one Watt.

No other processor architecture can offer implementations in these extremes; no other processor has ARM's broad applicability across an entire product range.

The *ARM Architecture Reference Manual* is the definitive description of the programmers' model of all ARM microprocessors, and is ARM's commitment to users of the ARM processor for compatibility and interworking across products designed and manufactured by many different companies.

Through this commitment, the benefits of an ARM processor can be harnessed to establish and maintain an industry lead that only ARM Powered products can achieve.

**Dave Jaggar. July 1996**

**ARM Architecture Reference Manual**
ARM DDI 0100B

# Contents

**ARM Architecture Reference Manual**

ARM DDI 0100B

# Contents

**ARM Architecture Reference Manual**
ARM DDI 0100B

# Contents

# Preface

This preface describes the ARM Architecture Version 4, also known as ARMv4, and lists the conventions and terminology used in this manual.

# Preface

## Introduction

The ARM Architecture exists in 4 major versions:

Version 1
: was implemented only by ARM1, and was never used in a commercial product.

Version 2
: extended Version 1 by adding a multiply instruction, multiply accumulate instruction, coprocessor support, two more banked registers for FIQ mode and later (Version 2a) an Atomic Load and Store instruction (called SWP) and the use of Coprocessor 15 as a system control coprocessor. Version 1, 2 and 2a all support a 26-bit address bus and combine in register 15 a 24-bit Program Counter (PC) and 8 bits of processor status.

Version 2 has just three privileged processing modes, Supervisor, IRQ and FIQ.

Version 3
: extended the addressing range to 32 bits, defining just a 30-bit Program Counter value in register 15, and added a separate 11-bit status register (the Current Program Status Register or CPSR) to contain the status information that was previously in register 15.

Version 3 added two new privileged processing modes to allow coprocessor emulation and virtual memory support in supervisor mode:

- Undefined
- Abort

Five more status registers (the Saved Program Status Registers, SPSRs) were defined, one for each privileged processor mode, to preserve the CPSR contents when the exception occurs corresponding to each privileged processor mode.

Version 3 supports both hardware and software emulation of Version 2a. The processor can be forced to be Version 2a only in hardware, allowing full backwards compatibility, but no version 3 advantages. Version 3 can also be switched in software into a Version 2 execution model to support Version 2a on a process by process basis to allow a smooth upgrade path from Version 2 to Version 3. See *Chapter 5, The 26-bit Architectures* for information on the differences between the 26-bit architectures (version1, 2 and 2a) and the 32-bit architectures (Version 3, 3M, 4 and 4T).

Version 3G
: is the same as version 3, without the backwards compatibility support for version 2 architectures.

Version 3M
: adds signed and unsigned multiply and multiply accumulate instructions that produce a 64-bit result to Version 3. The new ARMv3M instructions are marked as such in the text.

Version 4
: adds halfword load and store instructions, sign-extending byte and halfword load instructions, adds a new privileged processor mode (that uses the user mode registers) and defines several new undefined instructions.

Version 4T
: incorporates an instruction decoder for a 16-bit subset of the ARM instruction set (called THUMB).

**ARM Architecture Reference Manual**

ARM DDI 0100B

## Using this Manual

The information in this manual is organised as follows:

| | |
|---|---|
| **Chapter 1** | gives an overview of the ARM architecture |
| **Chapter 2** | describes the programmer's model |
| **Chapter 3** | lists and describes the ARM instruction set |
| **Chapter 4** | gives examples of coding algorithms |
| **Chapter 5** | describes the differences between 32-bit and 26-bit architectures |
| **Chapter 6** | lists and describes the Thumb instruction set |
| **Chapter 7** | describes ARM system architecture, and the system control processor |

## Conventions

This manual employs typographic conventions intended to improve its ease of use.

| | |
|---|---|
| `code` | a monospaced typewriter font shows code which you need to enter, or code which is provided as an example |

## Terminology

This manual uses the following terminology:

**Abort**
is caused by an illegal memory access. Aborts can be caused by the external memory system or the MMU

**AND**
performs a bitwise AND

**Arithmetic_Shift_Right**
performs a right shift, repeatedly inserting the original left-most bit (the sign bit) in the vacated bit positions on the left

**ARM instruction**
is a word that is word-aligned

**assert statements**
are used to ensure a certain condition has been met

**Assignment**
is signified by =

**Binary numbers**
are preceded by 0b

**Boolean AND**
is signified by the "and" operator

---

**ARM Architecture Reference Manual**
ARM DDI 0100B

**Boolean OR**
is signified by the "or" operator

**BorrowFrom**

returns 1    if the following subtract causes a borrow (the true unsigned result is less than 0)

returns 0    in all other cases

**Byte**
is an 8-bit data item

**CarryFrom**

returns 1    if the following addition causes a carry
(the result is bigger than $2^{31}-1$)

returns 0    in all other cases

**case ... endcase statements**
are used to indicate a one of many execution option. Indentation indicates the range of statements in each option

**Comments**
are enclosed in /* */

**ConditionPassed(cond)**

returns true    if the state of the N, Z, C and Z flags fulfils the condition encoded in the cond argument

returns false    in all other cases

**CPSR**
is the Current Program Status Register

**CurrentModeHasSPSR()**

returns true    if the current processor mode is not User mode or System mode

returns false    if the current mode is User mode or System mode

**Do-not-modify fields (DNM)**
means the value must not be altered by software. DNM fields read as
UNPREDICTABLE values, and may only be written with the same value read from the same field as the same processor.

**Elements**
are separated by I in a list of possible values for a variable

**EOR**
performs an Exclusive OR

**Exception**
is a mechanism to handle an event; for example, an external interrupt or an undefined instruction

**External abort**
is an abort that is generated by the external memory system

**ARM Architecture Reference Manual**

ARM DDI 0100B

**Fault**

is an abort that is generated by the MMU

**General-purpose register**

is one of the 32-bit general-purpose integer registers, R0 to R14

**Halfword**

is a 16-bit data item

**Hexadecimal numbers**

are preceded by 0x

**if ... else if ... else statements**

are used to signify conditional execution. Indentation indicates the range of statements in each condition

**IGNORE fields (IGN)**

must ignore writes

**Immediate and offset fields**

are unsigned unless otherwise stated

**IMPLEMENTATION-DEPENDENT fields (IMP)**

are not architecturally specified, but must be defined and documented by individual implementations

**InAPrivilegedMode()**

returns true if the current processor mode is not User mode;

returns false if the current mode is User mode

**Logical_Shift_Left**

performs a left shift, inserting zeroes in the vacated bit positions on the right. << is used as a short form for Logical_Shift_Left

**Logical_Shift_Right**

performs a right shift, inserting zeroes in the vacated bit positions on the left

**LR (Link Register)**

is integer register R14

**Memory[<address>,<size>]**

refers to a data item in memory of length <size>, at address <address>, aligned on a <size> byte boundary.
The data item is zero-extended to 32 bits.
Currently defined sizes are:

1 for bytes
2 for halfwords
4 for words

To align on a <size> boundary, halfword accesses ignore address[0] and word accesses ignore address[1:0]

**NOT**
>   performs a Complement

**NotFinished(CP_number)**
>   returns true      if the coprocessor signified by the CP_number argument has signalled that the current operation is complete
>
>   returns false     in all other cases

**NumberOfSetBitsIn(bitfield)**
>   performs a population count on (counts the set bits in) the bitfield argument

**Object[from:to]**
>   indicates the bit field extracted from Object starting at bit "from", ending with bit "to" (inclusive)

**Optional parts of instructions**
>   are surrounded by { and }

**OR**
>   performs an Inclusive OR

**OverflowFrom**
>   returns 1 if the following addition (or subtraction) causes a carry (or borrow) to (from) bit[31]. Addition generates a carry if both operands have the same sign (bit[31]), and the sign of the result is different to the sign of both operands. Subtraction causes an overflow if the operands have different signs, and the first operand and the result have different signs

**PC (Program Counter)**
>   is integer register R15 (or bits[25:2] of register 15 on 26-bit architectures)

**PSR**
>   is the CPSR or one of the SPSRs (or bits[31:26] and bits[1:0] of register 15 on 26-bit architectures)

**Read-as-zero fields (RAZ)**
>   appear as zero when read

**Read-Modify-Write fields (RMW)**
>   should be read to a general-purpose register, the relevant fields updated in the register, and the register value written back.

**Rotate_Right**
>   performs a right rotate, where each bit that is shifted off the right is inserted on the left

**Security hole**
>   is an illegal mechanism that bypasses system protection

**Should-be-one fields (SBO)**
>   should be written as one (or all 1s for bit fields) by software. Values other than one values produce unpredictable results

**ARM Architecture Reference Manual**

ARM DDI 0100B

**Should-be-one-or-preserved fields (SBOP)**
> should be written as one (or all 1s for bit fields) or preserved by writing the same value that has been previously read from the same fields on the same processor.

**Should-be-zero fields (SBZ)**
> should be written as zero (or all 0s for bit fields) by software.
> Non-zero values produce unpredictable results

**Should-be-zero-or-preserved fields (SBZP)**
> should be written as zero (or all 0s for bit fields) or preserved by writing the same value that has been previously read from the same fields on the same processor.

**Signed immediate and offset fields**
> are encoded in two's complement notation unless otherwise stated

**SignExtend(arg)**
> sign-extends (propagates the sign bit) its argument to 32 bits

**SPSR**
> is the Saved Program Status Register

**Test for equality**
> is signified by ==

**THUMB instruction**
> is a halfword that is halfword-aligned

**Unaffected items**
> are not changed by a particular operation

**UNDEFINED**
> indicates an instruction that generates an undefined instruction trap. See *2.5 Exceptions* on page 2-6 for information on undefined instruction traps

**UNPREDICTABLE**
> means the result of an instruction cannot be relied upon.
> unpredictable instructions or results must not represent security holes.
> **UNPREDICTABLE** instructions must not halt or hang the processor, or any parts of the system

**UNPREDICTABLE fields (UNP)**
> do not contain valid data, and a value may vary from moment to moment, instruction to instruction, and implementation to implementation

**Variable name**
> (a symbolic name for values) is surrounded by < and >

**while .... statements**
> are used to indicate a loop. Indentation indicates the range of statements in the loop

**Word**
> is a 32-bit data item

**1**

# Architecture Overview

# Architecture Overview 1

The ARM architecture has been designed to allow very small, yet high-performance implementations. It is the architectural simplicity of ARM which leads to very small implementations, and small implementations allow devices with very low power consumption.

# Architecture Overview

## 1.1 Overview

The ARM is a RISC (Reduced Instruction Set Computer), as it incorporates all the features of a typical RISC architecture:

- a large uniform register file
- a load-store architecture (data-processing operations only operate on register contents)
- simple addressing modes (data loaded and stored from an address specified in registers and instruction fields)
- uniform and fixed length instruction fields (which simplify instruction decode)

In addition, the ARM architecture provides these features:

- control over both the ALU and shifter in every data-processing instruction to maximise the use of a shifter and an ALU
- auto-increment and auto-decrement addressing modes to optimise program loops
- load and store multiple instructions to maximise data throughput
- conditional execution of all instructions to maximise execution throughput

Together, these architectural enhancements to a basic RISC architecture allow implementations that can balance high performance, low power consumption and minimal die size in every implementation.

### 1.1.1 ARM registers

ARM has thirty-one, 32-bit registers. At any one time, sixteen are visible; the other registers are used to speed up exception processing. All the register specifiers in ARM instructions can address any of the 16 registers.

The main bank of sixteen registers is used by all non-privileged code; these are the User mode registers. User mode is different from all other modes, as it is non-privileged, which means that user mode is the only mode which cannot switch to another processor mode (without generating an exception).

**Program counter**

Register 15 is the Program Counter (or PC), and can be used in most instructions as a pointer to the instruction which is two instructions after the instruction being executed. All ARM instructions are 4 bytes long (one 32-bit word), and are always aligned on a word boundary, so the PC contains just 30 bits; the bottom two bits are always zero.

**Link register**

Register 14 is called the Link Register (or LR). Its special purpose is to hold the address of the next instruction after a Branch with Link (BL) instruction, which is the instruction used to make a subroutine call. At all other times, R14 can be used as a general-purpose register.

**Other registers**

The remaining 14 registers have no special hardware purpose - their uses are defined purely by software. Software will normally use R13 as a stack pointer (or SP).

**ARM Architecture Reference Manual**

ARM DDI 0100B

## 1.2    Exceptions

ARM supports 5 types of exception, and a privileged processing mode for each type. The 5 types of exceptions are:

- two levels of interrupt (fast and normal)
- memory aborts (used to implement memory protection or virtual memory)
- attempted execution of an undefined instruction
- software interrupts (SWIs) (used to make a call to an Operating System)

When an exception occurs, some of the standard registers are replaced with registers specific to the exception mode. All exceptions have replacement (or banked) registers for R14 and R13, and one interrupt mode has more registers for fast interrupt processing.

After an exception, R14 holds the return address for exception processing, which is used both to return after the exception is processed and to address the instruction that caused the exception.

R13 is banked across exception modes to provide each exception handler with private stack pointer (SP). The fast interrupt mode also banks registers 8 to 12, so that interrupt processing can begin without the need to save or restore these registers. There is a seventh processing mode, System mode, that does not have any banked registers (it uses the User mode registers), which is used to run normal (non-exception) tasks that require a privileged processor mode.

### CPSR and SPSR

All other processor state is held in status registers. The current operating processor status is in the Current Program Status Register or CPSR. The CPSR holds:

- 4 condition code flags (Negative, Zero, Carry and Overflow)
- 2 interrupt disable bits (one for each type of interrupt)
- 5 bits which encode the current processor mode

All 5 exception modes also have a Saved Program Status Register (SPSR) which holds the CPSR of the task immediately before the exception occurred. Both the CPSR and SPSR are accessed with special instructions.

### The exception process

When an exception occurs, ARM halts execution after the current instruction and begins execution at a fixed address in low memory, known as the exception vectors. There is a separate vector location for each exception (and two for memory aborts to distinguish between data and instruction accesses).

An operating system will install an handler on every exception at initialisation. Privileged operating system tasks are normally run in System mode to allow exceptions to occur within the operating system without state loss (exceptions overwrite their R14 when an exception occurs, and System mode is the only privileged mode that cannot be entered by an exception).

## ARM Architecture Reference Manual

ARM DDI 0100B

# Architecture Overview

## 1.3  ARM Instruction Set

The ARM instruction set can be divided into four broad classes of instruction:

- branch
- data-processing
- load and store
- coprocessor

### Conditional execution

All ARM instructions may be conditionally executed. Data-processing instructions (and one type of coprocessor instruction) can update the four condition code flags in the CPSR (Negative, Zero, Carry and Overflow) according to their result. Subsequent instructions can be conditionally executed according to the status of the condition code flags. Fifteen conditions are implemented, depending on particular values of the condition code flags; one condition actually ignores the condition code flag so that normal (unconditional) instructions always execute.

## 1.4  Branch Instructions

As well as allowing any data-processing or load instruction to change control flow (by writing the Program Counter) a standard branch instruction is provided with 24-bit signed offset, allowing forward and backward branches of up to 32Mbytes.

There is a Branch with Link option that also preserves the address of the instruction after the branch in R14 (the Link Register or LR), allowing a move instruction to put the LR in to PC and return to the instruction after the branch, providing a subroutine call.

There also a special type of branch instruction called software interrupt (SWI). This makes a call to the Operating System (to request an OS-defined service). SWI also changes the processor mode, allowing an unprivileged task to gain OS privilege (access to which is controlled by the OS).

On processors that implement the THUMB instruction set there is a branch instruction that jumps to an address specified in a register, and optionally switches instruction set, allowing ARM code to call THUMB code and THUMB code to call ARM code. An overview of the THUMB instruction is provided in *Chapter 6, The Thumb Instruction Set*.

## 1.5  Data-processing Instructions

The data-processing instructions perform some operation on the general-purpose registers. There are three types of data-processing instructions:

- data-processing instructions proper
- multiply instructions
- status register transfer instructions

### Arithmetic/logic instructions

There are 16 arithmetic/logic instructions which share a common instruction format. This format takes up to two source operands, performs an arithmetic/logic operation on those operands, and then most store the result into a register, and optionally update the condition code flags according to that result.

**ARM Architecture Reference Manual**
ARM DDI 0100B

There are four data-processing instructions which don't store their result in a register. They compare the values of the source operands and then update the condition code flags.

Of the two source operands:

- one is always a register
- the other has two basic forms:
  - an immediate value
  - a register value, optionally shifted

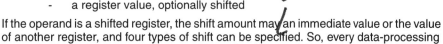

If the operand is a shifted register, the shift amount may an immediate value or the value of another register, and four types of shift can be specified. So, every data-processing instruction can perform a data-processing and a shift operation. As a result, ARM does not have dedicated shift instructions.

Because the Program Counter (PC) is a general-purpose register, this class of data-processing instruction may write their results directly to the PC, allowing a variety of jump instructions to be easily implemented.

### Multiply instructions

Multiply instructions come in two classes. Both types multiply the two 32-bit register values and store their result:

(normal) 32-bit result        store the 32-bit result in a register

(long) 64-bit result          store the 64 bit result in two separate registers

Both types of multiply instruction can optionally perform an accumulate operation.

### Status register transfer instructions

The status register transfer instructions transfer the contents of the CPSR or a SPSR to or from a general-purpose register. Writing to the CPSR is one way to set the value of the condition code flags and interrupt enable flags and to set the processor mode.

## 1.6    Load and Store Instructions

Load and store instruction come in three types:

1    load or store the value of a single register
2    load and store multiple register values
3    swap a register value with the value of a memory location

### Load and store single register

Load register instructions can load a 32-bit word, a 16-bit halfword or an 8-bit byte from memory into a register. Byte and halfword loads may be automatically zero- or sign-extended as they are loaded.

Store register instructions can store a 32-bit word, a 16-bit halfword or an 8-bit byte from a register to memory.

Load and store instructions have three primary addressing modes that are formed by adding or subtracting an immediate or register-based offset to or from a base register (register-based offsets may also be scaled with shift operations):

1. offset
2. pre-indexed
3. post-indexed

Pre- and post-indexed addressing modes update the base register with the base plus offset calculation. As the Program Counter (PC) is a general-purpose register, a 32-bit value can be loaded directly into the PC to perform a jump to any address in the 4Gbyte memory space.

### Load and store multiple registers

Load and Store multiple instructions perform a block transfer of any number of the general-purpose registers to or from memory. Four addressing modes are provided:

1. pre-increment
2. post-increment
3. pre-decrement
4. post-decrement

The base address is specified by a register value (which may be optionally updated after the transfer). As the subroutine return address and Program Counter (PC) values are in general-purpose registers, very efficient subroutine call and return can be constructed with Load and Store Multiple; register contents and the return address can be stacked and the stack pointer updated with single store multiple instruction at procedure entry and then register contents restored, the PC loaded with the return address and the stack pointer updated on procedure return with a single load multiple).

Of course, load and store multiple also allow very efficient data movement (for example, block copy).

### Swap a register value with the value of a memory location

Swap can load a value from a register-specified memory location, store the contents of a register to the same memory location, then write the loaded value to a register.

By specifying the same register as the load and store value, the contents of a memory location and a register are interchanged.

The swap operation performs a special indivisible bus operation that allows atomic update of semaphores. Both 32-bit word and 8-bit byte semaphores are supported.

## 1.7 Coprocessor Instructions

There are three types of coprocessor instructions:

| | |
|---|---|
| data-processing instructions | start a coprocessor-specific internal operation |
| register transfers | allow a coprocessor value to be transferred to or from an ARM register |
| data-transfer instructions | transfer coprocessor data to or from memory, where the address of the transfer is calculated by the ARM |

**ARM Architecture Reference Manual**

ARM DDI 0100B

**2**

# Programmer's Model

# 2 Programmer's Model

This chapter introduces the ARM Programmer's Model.

# Programmer's Model

## 2.1 Data Types

ARM Architecture Version 4 processors support the following data types:

| | |
|---|---|
| Byte | 8 bits |
| Halfword | 16 bits; halfwords must be aligned to two-byte boundaries |
| Word | 32 bits; words must be aligned to four-byte boundaries |

ARM instructions are exactly one word (and therefore aligned on a four-byte boundary).

THUMB instructions are exactly one halfword (and therefore aligned on a two-byte boundary).

All data operations (e.g. ADD) are performed on word quantities.

Load and store operations can transfer bytes, halfwords and words to and from memory, automatically zero-extending or sign-extending bytes or halfwords as they are loaded.

Signed operands are in two's complement format.

## 2.2 Processor Modes

ARM Version 4 supports seven processor modes:

| | Processor mode | | Description |
|---|---|---|---|
| 1 | User | (usr) | normal program execution mode |
| 2 | FIQ | (fiq) | supports a high-speed data transfer or channel process |
| 3 | IRQ | (irq) | used for general purpose interrupt handling |
| 4 | Supervisor | (svc) | a protected mode for the operating system |
| 5 | Abort | (abt) | implements virtual memory and/or memory protection |
| 6 | Undefined | (und) | supports software emulation of hardware coprocessors |
| 7 | System | (sys) | runs privileged operating system tasks (Architecture Version 4 only) |

*Table 2-1: ARM Version 4 processor modes*

Mode changes may be made under software control or may be caused by external interrupts or exception processing. Most application programs will execute in User mode. The other modes, known as privileged modes, will be entered to service interrupts or exceptions or to access protected resources.

**ARM Architecture Reference Manual**
ARM DDI 0100B

## 2.3    Registers

The processor has a total of 37 registers:

- 30 general-purpose registers
- 6 status registers
- a program counter

The registers are arranged in partially overlapping banks: a different register bank for each processor mode. At any one time, 15 general-purpose registers (R0 to R14), one or two status registers and the program counter are visible. The general-purpose registers and status registers currently visible depend on the current processor mode. The register bank organisation is shown in *Figure 2-1: Register organisation* on page 2-4. The banked registers are shaded in the diagram.

The general-purpose registers are 32 bits wide.

Register 13 (the Stack Pointer or SP) is banked across all modes to provide a private Stack Pointer for each mode (except system mode which shares the user mode R13).

Register 14 (the Link Register or LR) is used as the subroutine return address link register. R14 is also banked across all modes (except system mode which shares the user mode R14). When a Subroutine call (Branch and Link instruction) is executed, R14 is set to the subroutine return address; R14_svc, R14_irq, R14_fiq, R14_abort and R14_undef are used similarly to hold the return address when exceptions occur (or a subroutine return address if subroutine calls are executed within interrupt or exception routines). R14 may be treated as a general-purpose register at all other times.

FIQ mode also has banked registers R8 to R12 (as well as R13 and R14). R8_fiq, R9_fiq, R10_fiq, R11__fiq and R12_fiq are provided to allow very fast interrupt processing (without the need to preserve register contents by storing them to memory), and to preserve values across interrupt calls (so that register contents do not need to be restored from memory).

Register R15 holds the Program Counter (PC). When R15 is read, bits [1:0] are zero and bits [31:2] contain the PC. When R15 is written bits[1:0] are ignored and bits[31:2] are written to the PC. Depending on how it is used, the value of the PC is either the address of the instruction plus 8 or is UNPREDICTABLE.

The Current Program Status Register (CPSR) is also accessible in all processor modes. It contains condition code flags, interrupt enable flags and the current mode. Each privileged mode (except system mode) has a Saved Program Status Register (SPSR) which is used to preserve the value of the CPSR when an exception occurs.

## 2.4    Program Status Registers

The format of the Current Program Status Register (CPSR) and the Saved Program Status registers (SPSR) are shown in *Figure 2-2: Format of the program status registers* on page 2-4. The N, Z, C and V (Negative, Zero, Carry and oVerflow) bits are collectively known as the condition code flags. The condition code flags in the CPSR can be changed as a result of arithmetic and logical operations in the processor and can be tested by all instructions to determine if the instruction is to be executed.

**ARM Architecture Reference Manual**
ARM  DDI 0100B

# Programmer's Model

| Mode | | | | | |
|---|---|---|---|---|---|
| **User/System** | **Supervisor** | **Abort** | **Undefined** | **Interrupt** | **Fast interrupt** |
| R0 | R0 | R0 | R0 | R0 | R0 |
| R1 | R1 | R1 | R1 | R1 | R1 |
| R2 | R2 | R2 | R2 | R2 | R2 |
| R3 | R3 | R3 | R3 | R3 | R3 |
| R4 | R4 | R4 | R4 | R4 | R4 |
| R5 | R5 | R5 | R5 | R5 | R5 |
| R6 | R6 | R6 | R6 | R6 | R6 |
| R7 | R7 | R7 | R7 | R7 | R7 |
| R8 | R8 | R8 | R8 | R8 | R8_FIQ |
| R9 | R9 | R9 | R9 | R9 | R9_FIQ |
| R10 | R10 | R10 | R10 | R10 | R10_FIQ |
| R11 | R11 | R11 | R11 | R11 | R11_FIQ |
| R12 | R12 | R12 | R12 | R12 | R12_FIQ |
| R13 | R13_SVC | R13_ABORT | R13_UNDEF | R13_IRQ | R13_FIQ |
| R14 | R14_SVC | R14_ABORT | R14_UNDEF | R14_IRQ | R14_FIQ |
| PC | PC | PC | PC | PC | PC |

| | | | | | |
|---|---|---|---|---|---|
| CPSR | CPSR | CPSR | CPSR | CPSR | CPSR |
| | SPSR_SVC | SPSR_ABORT | SPSR_UNDEF | SPSR_IRQ | SPSR_FIQ |

= banked register

*Figure 2-1: Register organisation*

31 30 29 28 27 26 25 24 23 22 21 20 19 18 17 16 15 14 13 12 11 10 9 8 7 6 5 4 3 2 1 0

| N | Z | C | V | Q | DNM/RAZ | I | F | T | M4 | M3 | M2 | M1 | M0 |
|---|---|---|---|---|---|---|---|---|---|---|---|---|---|

*Figure 2-2: Format of the program status registers*

**ARM Architecture Reference Manual**
ARM DDI 0100B

## 2.4.1　The control bits

The bottom 8 bits of a PSR (incorporating I, F, T and M[4:0]) are known collectively as the **control bits.** The control bits change when an exception arises and can be altered by software only when the processor is in a privileged mode. The I and F bits are the interrupt disable bits:

I bit　　　　disables IRQ interrupts when it is set

F bit　　　　disables FIQ interrupts when it is set

The T flag is only implemented on Architecture Version 4T (THUMB):

0　　　　　　indicates ARM execution

1　　　　　　indicates THUMB execution

On all other version of the architecture the T flag should be zero (SBZ).

## 2.4.2　The mode bits

The M0, M1, M2, M3 and M4 bits (M[4:0]) are the mode bits, and these determine the mode in which the processor operates. The interpretation of the mode bits is shown in *Table 2-2: The mode bits.* Not all combinations of the mode bits define a valid processor mode. Only those explicitly described can be used; if any other value is programmed into the mode bits M[4:0], the result is unpredictable.

| M[4:0] | Mode | Accessible Registers |
|--------|------|----------------------|
| 0b10000 | User | PC, R14 to R0, CPSR |
| 0b10001 | FIQ | PC, R14_fiq to R8_fiq, R7 to R0, CPSR, SPSR_fiq |
| 0b10010 | IRQ | PC, R14_irq, R13_irq, R12 to R0, CPSR, SPSR_irq |
| 0b10011 | SVC | PC, R14_svc, R13_svc, R12 to R0, CPSR, SPSR_svc |
| 0b10111 | Abort | PC, R14_abt, R13_abt, R12 to R0, CPSR, SPSR_abt |
| 0b11011 | Undef | PC, R14_und, R13_und, R12 to R0, CPSR, SPSR_und |
| 0b11111 | System | PC, R14 to R0, CPSR (Architecture Version 4 only) |

*Table 2-2: The mode bits*

User mode and system mode do not have an SPSR, as these modes are not entered on any exception, so a register to preserve the CPSR is not required. In User mode or System mode any reads to the SPSR will read an unpredictable value, and any writes to the SPSR will be ignored.

## 2.5  Exceptions

Exceptions are generated by internal and external sources to cause the processor to handle an event; for example, an externally generated interrupt, or an attempt to execute an undefined instruction. The processor state just before handling the exception must be preserved so that the original program can be resumed when the exception routine has completed. More than one exception may arise at the same time.

ARM supports 7 types of exception and has a privileged processor mode for each type of exception. *Table 2-3: Exception processing modes* lists the types of exception and the processor mode that is used to process that exception. When an exception occurs execution is forced from a fixed memory address corresponding to the type of exception. These fixed addresses are called the **Hard Vectors.**

The reserved entry at address 0x14 is for an Address Exception vector used when the processor is configured for a 26-bit address space. See *Chapter 5, The 26-bit Architectures* for more information.

| Exception type | Mode | Vector address |
|---|---|---|
| Reset | SVC | 0x00000000 |
| Undefined instructions | UNDEF | 0x00000004 |
| Software Interrupt (SWI) | SVC | 0x00000008 |
| Prefetch Abort (Instruction fetch memory abort) | ABORT | 0x0000000c |
| Data Abort (Data Access memory abort) | ABORT | 0x00000010 |
| IRQ (Interrupt) | IRQ | 0x00000018 |
| FIQ (Fast Interrupt) | FIQ | 0x0000001c |

*Table 2-3: Exception processing modes*

When taking an exception, the banked registers are used to save state. When an exception occurs, these actions are performed:

```
R14_<exception_mode> = PC
SPSR_<exception_mode> = CPSR
CPSR[5:0] = Exception mode number
CPSR[6] = if <exception_mode> == Reset or FIQ then = 1 else unchanged
CPSR[7] = 1; Interrupt disabled
PC = Exception vector address
```

To return after handling the exception, the SPSR is moved into the CPSR and R14 is moved to the PC. This can be done atomically in two ways:

1   Using a data-processing instruction with the S bit set, and the PC as the destination.

2   Using the Load Multiple and Restore PSR instruction.

The following sections show the recommended way of returning from each exception.

**ARM Architecture Reference Manual**
ARM  DDI 0100B

## 2.5.1 Reset

When the processor's Reset input is asserted, ARM immediately stops execution of the current instruction. When the Reset is de-asserted, the following actions are performed:

```
R14_svc = unpredictable value
SPSR_svc = CPSR
CPSR[5:0] = 0b010011    ; Supervisor mode
CPSR[6] = 1             ; Fast Interrupts disabled
CPSR[7] = 1             ; Interrupts disabled
PC = 0x0
```

Therefore, after reset, ARM begins execution at address 0x0 in supervisor mode with interrupts disabled. See *7.6 Memory Management Unit (MMU) Architecture* on page 7-14 for more information on the effects of Reset.

## 2.5.2 Undefined instruction exception

If ARM executes a coprocessor instruction, it waits for any external coprocessor to acknowledge that it can execute the instruction. If no coprocessor responds, an undefined instruction exception occurs. If an attempt is made to execute an instruction that is undefined, an undefined instruction exception occurs (see *3.14.5 Undefined instruction Space* on page 3-27).

The undefined instruction exception may be used for software emulation of a coprocessor in a system that does not have the physical coprocessor (hardware), or for general-purpose instruction set extension by software emulation.

When an undefined instruction exception occurs, the following actions are performed:

```
R14_und = address of undefined instruction + 4
SPSR_und = CPSR
CPSR[5:0] = 0b011011    ; Undefined mode
CPSR[6] = unchanged     ; Fast Interrupt status is unchanged
CPSR[7] = 1             ; (Normal) Interrupts disabled
PC = 0x4
```

To return after emulating the undefined instruction, use:

```
MOVS PC,R14
```

This restores the PC (from R14_und) and CPSR (from SPSR_und) and returns to the instruction following the undefined instruction.

## 2.5.3 Software interrupt exception

The software interrupt instruction (SWI) enters Supervisor mode to request a particular supervisor (Operating System) function. When a SWI is executed, the following are performed:

```
R14_svc = address of SWI instruction + 4
SPSR_svc = CPSR
CPSR[5:0] = 0b010011    ; Supervisor mode
CPSR[6] = unchanged     ; Fast Interrupt status is unchanged
CPSR[7] = 1             ; (Normal) Interrupts disabled
PC = 0x8
```

To return after performing the SWI operation, use:

```
MOVS PC,R14
```

This restores the PC (from R14_svc) and CPSR (from SPSR_svc) and returns to the instruction following the SWI.

## 2.5.4 Prefetch Abort (Instruction Fetch Memory Abort)

A memory abort is signalled by the memory system. Activating an abort in response to an instruction fetch marks the fetched instruction as invalid. An abort will take place if the processor attempts to execute the invalid instruction. If the instruction is not executed (for example as a result of a branch being taken while it is in the pipeline), no prefetch abort will occur.

When an attempt is made to execute an aborted instruction, the following actions are performed:

```
R14_abt = address of the aborted instruction + 4
SPSR_abt = CPSR
CPSR[5:0] = 0b010111    ; Abort mode
CPSR[6] = unchanged     ; Fast Interrupt status is unchanged
CPSR[7] = 1             ; (Normal) Interrupts disabled
PC = 0xc
```

To return after fixing the reason for the abort, use:

```
SUBS PC,R14,#4
```

This restores both the PC (from R14_abt) and CPSR (from SPSR_abt) and returns to the aborted instruction.

## 2.5.5 Data Abort (Data Access Memory Abort)

A memory abort is signalled by the memory system. Activating an abort in response to a data access (Load or Store) marks the data as invalid. A data abort exception will occur before any following instructions or exceptions have altered the state of the CPU, and the following actions are performed:

```
R14_abt = address of the aborted instruction + 8
SPSR_abt = CPSR
CPSR[5:0] = 0b010111    ; Abort mode
CPSR[6] = unchanged     ; Fast Interrupt status is unchanged
CPSR[7] = 1             ; (Normal) Interrupts disabled
PC = 0x10
```

To return after fixing the reason for the abort, use:

```
SUBS PC,R14,#8
```

This restores both the PC (from R14_abt) and CPSR (from SPSR_abt) and returns to re-execute the aborted instruction.

If the aborted instruction does not need to be re-executed use:

*ARM code only*

```
SUBS PC,R14,#4
```

*#6 in Thumb mode*

The final value left in the base register used in memory access instructions which specify writeback and generate a data abort (LDR, LDRH, LDRSH, LDRB, LDRSB, STR, STRH, STRB, LDM, STM, LDC, STC) is IMPLEMENTATION DEFINED.

An implementation can choose to leave either the original value or the updated value in the base register, but the same behaviour must be implemented for all memory access instructions.

## 2.5.6 IRQ (Interrupt Request) exception

The IRQ (Interrupt ReQuest) exception is externally generated by asserting the processor's IRQ input. It has a lower priority than FIQ (see below), and is masked out when a FIQ sequence is entered. Interrupts are disabled when the I bit in the CPSR is set (but note that the I bit can only be altered from a privileged mode). If the I flag is clear, ARM checks for a IRQ at instruction boundaries.

When an IRQ is detected, the following actions are performed:

```
R14_irq = address of next instruction to be executed + 4
SPSR_irq = CPSR
CPSR[5:0] = 0b010010    ; Interrupt mode
CPSR[6] = unchanged     ; Fast Interrupt status is unchanged
CPSR[7] = 1             ; (Normal) Interrupts disabled
PC = 0x18
```

To return after servicing the interrupt, use:

```
SUBS PC,R14,#4
```

This restores both the PC (from R14_irq) and CPSR (from SPSR_irq) and resumes execution of the interrupted code.

## 2.5.7 FIQ (Fast Interrupt Request) exception

The FIQ (Fast Interrupt reQuest) exception is externally generated by asserting the processor's FIQ input. FIQ is designed to support a data transfer or channel process, and has sufficient private registers to remove the need for register saving in such applications (thus minimising the overhead of context switching).

Fast interrupts are disabled when the F bit in the CPSR is set (but note that the F bit can only be altered from a privileged mode). If the F flag is clear, ARM checks for a FIQ at instruction boundaries.

When a FIQ is detected, the following actions are performed:

```
R14_fiq = address of next instruction to be executed + 4
SPSR_fiq = CPSR
CPSR[5:0] = 0b010001    ; FIQ mode
CPSR[6] = unchanged     ; Fast Interrupt disabled
CPSR[7] = 1             ; Interrupts disabled
PC = 0x1c
```

To return after servicing the interrupt, use:

```
SUBS PC, R14, #4
```

This restores both the PC (from R14_fiq) and CPSR (from SPSR_fiq) and resumes execution of the interrupted code.

The FIQ vector is deliberately the last vector to allow the FIQ exception-handler software to be placed directly at address 0x1c, and not require a branch instruction from the vector.

## 2.5.8 Exception priorities

The Reset exception has the highest priority. FIQ has higher priority than IRQ. IRQ has higher priority than prefetch abort.

Undefined instruction and software interrupt cannot occur at the same time, as they each correspond to particular (non-overlapping) decodings of the current instruction, and both must be lower priority than prefetch abort, as a prefetch abort indicates that no valid instruction was fetched.

The priority of data abort is higher than FIQ and lower priority than Reset, which ensures that the data-abort handler is entered before the FIQ handler is entered (so that the data abort will be resolved after the FIQ handler has completed).

| Exception | Priority |
|---|---|
| Reset | 1 (Highest) |
| Data Abort | 2 |
| FIQ | 3 |
| IRQ | 4 |
| Prefetch Abort | 5 |
| Undefined Instruction, SWI | 6 (Lowest) |

*Table 2-4: Exception priorities*

**ARM Architecture Reference Manual**
ARM DDI 0100B

**3**

# The ARM Instruction Set

# The ARM Instruction Set

This chapter describes the ARM instruction set.

# The ARM Instruction Set

## 3.1    Using this Chapter

This chapter is divided into three parts:

1    Overview of the ARM instruction types
2    Alphabetical list of instructions
3    Addressing modes

### 3.1.1    Overview of the ARM instruction types *(page 3-3 through 3-27)*

This part describes the functional groups within the instruction set, and shows relevant examples and encodings. Each functional group lists all its instructions, which you can then find in the alphabetical section. The functional groups are:

1    Branch
2    Data processing
3    Multiply
4    Status register access
5    Load and store:
  -    load and store word or unsigned byte
  -    load and store halfword and load signed byte
  -    load and store multiple
6    Semaphore
7    Coprocessor

### 3.1.2    Alphabetical list of instructions *(page 3-30 through 3-81)*

This part lists every ARM instruction, and gives:

• instruction syntax and functional group
• encoding and operation
• relevant exceptions and qualifiers
• notes on usage
• restrictions on availability in versions of the ARM architecture
• a cross-reference to the relevant addressing modes

### 3.1.3    Addressing modes *(page 3-84 through 3-126)*

This part lists the addressing modes for the functional groups of instructions:

| | |
|---|---|
| Mode 1 | Shifter operands for data processing instuctions |
| Mode 2 | Load and store word or unsigned byte addressing modes |
| Mode 3 | Load and store halfword or load signed byte addressing modes |
| Mode 4 | Load and store multiple addressing modes |
| Mode 5 | Load and store coprocessor addressing modes |

**ARM Architecture Reference Manual**
ARM DUI 0100B

## 3.2　Instruction Set Overview

*Table 3-1: ARM instruction set overview (expanded)* shows the instruction set encoding. All other bit patterns are UNPREDICTABLE.

| | 31 30 29 28 | 27 26 25 | 24 23 22 21 | 20 | 19 18 17 16 | 15 14 13 12 | 11 10 9 8 | 7 6 5 | 4 | 3 2 1 0 |
|---|---|---|---|---|---|---|---|---|---|---|
| Data processing Immediate | cond | 0 0 1 | op | S | Rn | Rd | rotate | immediate | | |
| Data processing Immediate shift | cond | 0 0 0 | opcode | S | Rn | Rd | shift immed | shift | 0 | Rm |
| Data processing register shift | cond | 0 0 0 | opcode | S | Rn | Rd | Rs | 0 shift | 1 | Rm |
| Multiply | cond | 0 0 0 | 0 0 0 A | S | Rd | Rn | Rs | 1 0 0 | 1 | Rm |
| Multiply long | cond | 0 0 0 | 0 1 U A | S | RdHi | RdLo | Rs | 1 0 0 | 1 | Rm |
| Move from Status register *MRS* | cond | 0 0 0 | 1 0 R 0 | 0 | SBO | Rd | SBZ | 0 SBZ | | |
| Move immediate to Status register *msr* | cond | 0 0 1 | 1 0 R 1 | 0 | Mask | SBO | rotate | immediate | | |
| Move register to Status register *msr* | cond | 0 0 0 | 1 0 R 1 | 0 | Mask | SBO | SBZ | 0 | | Rm |
| Branch/Exchange instruction set | cond | 0 0 0 | 1 0 0 1 | 0 | SBO | SBO | SBO | 0 0 0 | 1 | Rm |
| Load/Store immediate offset | cond | 0 1 0 | P U B W | L | Rn | Rd | immediate | | | |
| Load/Store register offset | cond | 0 1 1 | P U B W | L | Rn | Rd | shift immed | shift | 0 | Rm |
| Load/Store halfword/signed byte | cond | 0 0 0 | P U 1 W | L | Rn | Rd | Hi Offset | 1 S H | 1 | Lo Offset |
| Load/Store halfword/signed byte | cond | 0 0 0 | P U 0 W | L | Rn | Rd | SBZ | 1 S H | 1 | Rm |
| Swap/Swap byte | cond | 0 0 0 | 1 0 B 0 | 0 | Rn | Rd | SBZ | 1 0 0 | 1 | Rm |
| Load/Store multiple | cond | 1 0 0 | P U S W | L | Rn | Register List | | | | |
| Coprocessor data processing | cond | 1 1 1 | 0 op1 | | CRn | CRd | cp_num | op2 | 0 | CRm |
| Coprocessor register transfers | cond | 1 1 1 | 0 op1 | L | CRn | Rd | cp_num | op2 | 1 | CRm |
| Coprocessor load and store | cond | 1 1 0 | P U N W | L | Rn | CRd | cp_num | 8_bit_offset | | |
| Branch and Branch with link | cond | 1 0 1 | L | | 24_bit_offset | | | | | |
| Software interrupt | cond | 1 1 1 | 1 | | swi_number | | | | | |
| Undefined instruction | cond | 0 1 1 | x x x x | x | x x x x | x x x x | x x x x | x x x | 1 | x x x x |

**Table 3-1: ARM instruction set overview (expanded)**

# The ARM Instruction Set

## 3.3 The Condition Field

All ARM instructions can be conditionally executed, which means that their execution may or may not take place depending on the values of values of the N, Z, C and V flags in the CPSR. Every instruction contains a 4-bit condition code field in bits 31 to 28, as shown in *Figure 3-1: Condition code fields*.

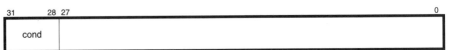

*Figure 3-1: Condition code fields*

### 3.3.1 Condition codes

This field specifies one of 16 conditions as described in *Table 3-2: Condition codes* on page 3-5. Every instruction mnemonic may be extended with the letters defined in the mnemonic extension field.

If the always (AL) condition is specified, the instruction will be executed irrespective of the value of the condition code flags. Any instruction that uses the never (NV) condition is UNPREDICTABLE. The absence of a condition code on an instruction mnemonic implies the always (AL) condition code.

**ARM Architecture Reference Manual**
ARM DUI 0100B

| Opcode [31:28] | Mnemonic Extension | Meaning | Status flag state |
|---|---|---|---|
| 0000 | EQ | Equal | Z set |
| 0001 | NE | Not Equal | Z clear |
| 0010 | CS/HS | Carry Set /Unsigned Higher or Same | C set |
| 0011 | CC/LO | Carry Clear /Unsigned Lower | C clear |
| 0100 | MI | Minus / Negative | N set |
| 0101 | PL | Plus /Positive or Zero | N clear |
| 0110 | VS | Overflow | V set |
| 0111 | VC | No Overflow | V clear |
| 1000 | HI | Unsigned Higher | C set and Z clear |
| 1001 | LS | Unsigned Lower or Same | C clear or Z set |
| 1010 | GE | Signed Greater Than or Equal | N set and V set, or N clear and V clear (N = V) |
| 1011 | LT | Signed Less Than | N set and V clear, or N clear and V set (N != V) |
| 1100 | GT | Signed Greater Than | Z clear, and either N set and V set, or N clear and V clear (Z = 0,N = V) |
| 1101 | LE | Signed Less Than or Equal | Z set, or N set and V clear, or N clear and V set (Z = 1, N != V) |
| 1110 | AL | Always (unconditional) | - |
| 1111 | NV | Never | - |

*Table 3-2: Condition codes*

# The ARM Instruction Set

## 3.4    Branch Instructions

All ARM processors support a branch instruction that allows a conditional branch forwards or backwards up to 32 Mbytes. As the Program Counter (PC) is one of the general-purpose registers (register 15), a branch or jump can also be generated by writing a value to register 15.

A subroutine call is a variant of the standard branch; as well as allowing a branch forward or backward up to 32 Mbytes, the Branch with Link instruction preserves the address of the instruction after the branch (the return address) in register 14 (the link register or LR).

Lastly, a load instruction provides a way to branch any where in the 4Gbyte address space (known as a long branch). A 32-bit value is loaded directly from memory into the PC, causing a branch. The load instruction may be preceded with a Move instruction to store a return address in the link register (LR or R14)).

### Examples

```
        B     label     ; branch unconditionally to label
        BCC   label     ; branch to label if carry flag is clear
        BEQ   label     ; branch to label if zero flag is set

        MOV   PC, #0     ; R15 = 0, branch to location zero

        BL    func       ; subroutine call to function

func    .

        .

        MOV   PC, LR     ; R15=R14, return to instruction after the BL

        MOV   LR, PC     ; store the address of the instruction after
                         ; the next one into R14 ready to return
        LDR   PC, =func  ; load a 32 bit value into the program counter
        .
```

Processors that support the Thumb instruction set (Architecture v4T) also support a branch instruction (BX) that jumps to a given address, and optionally switches executing Thumb instructions.

### 3.4.1    List of branch instructions

**ARM Architecture Reference Manual**
ARM DUI 0100B

## 3.5    Data Processing

ARM has 16 data-processing instructions. Most data-processing instructions take two source operands (Move and Move Not have only one operand) and store a result in a register (except for the Compare and Test instructions which only update the condition codes). Of the two source operands, one is always a register, the other is called a shifter operand, and is either an immediate value or a register. If the second operand is a register value, it may have a shift applied to it before it is used as the operand to the ALU.

| Mnemonic | Operation | Opcode | Action |
| --- | --- | --- | --- |
| MOV | Move | 1101 | Rd := shifter_operand (no first operand) |
| MVN | Move Not | 1111 | Rd := NOT shifter_operand (no first operand) |
| ADD | Add | 0100 | Rd := Rn + shifter_operand |
| ADC | Add with Carry | 0101 | Rd := Rn + shifter_operand + Carry Flag |
| SUB | Subtract | 0010 | Rd := Rn - shifter_operand |
| SBC | Subtract with Carry | 0110 | Rd := Rn - shifter_operand - NOT(Carry Flag) |
| RSB | Reverse Subtract | 0011 | Rd := shifter_operand - Rn |
| RSC | Reverse Subtract with Carry | 0111 | Rd := shifter_operand - Rn - NOT(Carry Flag) |
| AND | Logical AND | 0000 | Rd := Rn AND shifter_operand |
| EOR | Logical Exclusive OR | 0001 | Rd := Rn EOR shifter_operand |
| ORR | Logical (inclusive) OR | 1100 | Rd := Rn OR shifter_operand |
| BIC | Bit Clear | 1110 | Rd := Rn AND NOT shifter_operand |
| CMP | Compare | 1010 | update flags after Rn - shifter_operand |
| CMN | Compare Negated | 1011 | update flags after Rn + shifter_operand |
| TST | Test | 1000 | update flags after Rn AND shifter_operand |
| TEQ | Test Equivalence | 1001 | update flags after Rn EOR shifter_operand |

*Table 3-3: Data-processing instructions*

# The ARM Instruction Set

## 3.5.1 Instruction encoding

```
<opcode1>{<cond>}{S}  Rd, <shifter_operand>
<opcode1> := MOV | MVN


<opcode2>{<cond>}  Rn, <shifter_operand>
<opcode2> := CMP | CMN | TST | TEQ


<opcode3>{<cond>}{S}  Rd, Rn, <shifter_operand>
<opcode3> := ADD|SUB|RSB|ADC|SBC|RSC|AND|BIC|EOR|ORR
```

| 31 | 28 | 27 26 | 25 | 24 | 21 | 20 | 19 | 16 | 15 | 12 | 11 | 0 |
|---|---|---|---|---|---|---|---|---|---|---|---|---|
| cond | | 0 0 | I | opcode | | S | Rn | | Rd | | shifter_operand | |

Rd                          specifies the destination register

Rn                          specifies the first source operand register

`<shifter_operand>`         specifies the second source operand.
                            See *3.16 Data-processing Operands* on
                            page 3-84 for details of the shifter operands.

**Notes**

**The I bit:** Bit 25 is used to distinguish between the immediate and register forms of
`<shifter_operand>`.

**The S bit:** Bit 20 is used to signify that the instruction updates the condition codes.

## 3.5.2 Condition code flags

Data-processing instructions can update the four condition code flags.

CMP, CMN, TST and TEQ always update the condition code flags; the remaining
instructions will update the flags if an S is appended to the instruction mnemonic
(which sets the S bit in the instruction).

Bits are set as follows:

N (Negative) flag    is set if the result of a data-processing instruction is
                     negative

Z (Zero) flag        is set if the result is equal to zero

C (Carry) flag       is set if an add, subtract or compare causes a carry
                     (result bigger than 32 bits), or is set from the output of
                     the shifter for move and logical instructions

V (Overflow) flag    is set if an Add or Subtract, or compare overflows
                     (signed result bigger than 31 bits); unaffected by move or
                     ~~conditional~~ instructions

*logical*

**ARM Architecture Reference Manual**
ARM DUI 0100B

## 3.5.3    List of data-processing instructions

# The ARM Instruction Set

## 3.6 Multiply Instructions

ARM has two classes of multiply instruction:

- normal, 32-bit result
- long, 64-bit result

All multiply instructions take two register operands as the input to the multiplier. ARM does not directly support a multiply-by-constant instruction due to the efficiency of shift and add, or shift and reverse subtract instructions.

### 3.6.1 Normal multiply

There are two multiply instructions that produce 32-bit results:

MUL    multiplies the values of two registers together, truncates the result to 32 bits, and stores the result in a third register.

MLA    multiplies the values of two registers together, adds the value of a third register, truncates the result to 32 bits, and stores the result into a fourth register (i.e. performs multiply and accumulate).

Both multiply instructions can optionally set the N (Negative) and Z (Zero) condition code flags. No distinction is made between signed and unsigned variants; only the least-significant 32 bits of the result are stored in the destination register, and the sign of the operands does not affect this value.

```
MUL    R4, R2, R1      ; Set R4 to value of R2 multiplied by R1
MULS   R4, R2, R1      ; R4 = R2 x R1, set N and Z flags
MLA    R7, R8, R9, R3  ; R7 = R8 x R9 + R3
```

### 3.6.2 Long multiply

There are four multiply instructions that produce 64-bit results (long multiply).

Two of the variants multiply the values of two registers together and store the 64-bit result in a third and fourth register. There are a signed (SMULL) and unsigned (UMULL) variants. (The signed variants produce a different result in the most significant 32 bits if either or both of the source operands is negative).

The remaining two variants multiply the values of two registers together, add the 64-bit value from and third and fourth register and store the 64-bit result back into those registers (third and fourth). There are also signed (SMLAL) and unsigned (UMLAL) variants. These instruction perform a long multiply and accumulate.

All four long multiply instructions can optionally set the N (Negative) and Z (Zero) condition code flags:

```
SMULL  R4, R8, R2, R3  ; R4 = bits 0 to 31 of R2 x R3
                       ; R8 = bits 32 to 63 of R2 x R3
UMULL  R6, R8, R0, R1  ; R6, R8 = R0 x R1
UMLAL  R5, R8, R0, R1  ; R5, R8 = R0 x R1 + R5, R8
```

**ARM Architecture Reference Manual**
ARM DUI 0100B

### 3.6.3     List of multiply instructions

# The ARM Instruction Set

## 3.7    Status Register Access

There are two instructions for moving the contents of a program status register to or from a general-purpose register. Both the CPSR and SPSR can be accessed. Each status register is split into four 8-bit fields than can be individually written:

| bits 31 to 24 | the flags field |
| bits 23 to 16 | the status field |
| bits 15 to 8 | the extension field |
| bits 7 to 0 | the control field |

ARMv4 does not use the status and extension field, and 4 bits are unused in the flags field. The four condition code flags occupy the remaining four bits of the flags field, and the control field contains two interrupt disable bits, 5 processor mode bits, and the Thumb bit on ARMv4T (See *2.4 Program Status Registers* on page 2-3).

The unused bits of the status registers may be used in future ARM architectures, and should not be modified by software. Therefore, a read-modify write strategy should be used to update the value of a status register to ensure future compatibility. The status registers are readable to allow the read part of the read-modify-write operation, and to allow all processor state to be preserved (for instance, during process context switches). The status registers are writeable to allow the write part of the read-modify-write operation, and allow all processor state to be restored.

### 3.7.1    CPSR value

Altering the value of the CPSR has three uses:

1    Sets the value of the condition code flags to a known value.

2    Enables or disables interrupts.

3    Changes processor mode (for instance to initialise stack pointers).

### 3.7.2    Examples

```
MRS    R0, CPSR              ; Read the CPSR
BIC    R0, R0, #0xf0000000 ; Clear the N, Z, C and V bits
MSR    CPSR_f, R0            ; update the flag bits in the CPSR
; N, Z, C and V flags now all clear
MRS    R0, CPSR              ; Read the CPSR
ORR    R0, R0, #0x80         ; Set the interrupt disable bit
MSR    CPSR_c, R0            ; Update the control bits in the CPSR
; interrupts (IRQ) now disabled
MRS    R0, CPSR              ; Read the CPSR
BIC    R0, R0, #0x1f         ; Clear the mode bits
ORR    R0, R0, #0x11         ; Set the mode bits to FIQ mode
MSR    CPSR_c, R0            ; Update the control bits in the CPSR
; now in FIQ mode
```

### 3.7.3    List of status register access instructions

**ARM Architecture Reference Manual**

ARM DUI 0100B

## 3.8    Load and Store Instructions

ARMv4 supports two broad classes of instruction which load or store the value of a single register from or to memory.

- The first form can load or store a 32-bit word or an 8-bit unsigned byte
- The second form can load or store a 16-bit unsigned halfword, and can load and sign extend a 16-bit halfword or an 8-bit byte

The first form (word and unsigned byte) allows a wider range of addressing modes the second (halfword and signed byte). The Word and Unsigned Byte addressing mode comes in two parts:

- the base register
- the offset

The base register is any one of the general-purpose registers (including the PC, which allows PC-relative addressing for position-independent code).

The offset takes one of three forms:

1    Immediate Offset:
The offset is a 12-bit unsigned number, that may be added to or subtracted from the base register. Immediate Offset addressing is useful for accessing data elements that are a fixed distance from the start of the data object, such as structure fields, stack offsets and IO registers.

2    Register Offset:
The offset is a general-purpose register (not the PC), that may be added to or subtracted from the base register. Register offsets are useful for accessing arrays or blocks of data.

3    Scaled Register Offset:
The offset is a general-purpose register (not the PC) shifted by an immediate value, then added to or subtracted from the base register. The same shift operations used for data-processing instructions can be used (Logical Shift Left, Logical Shift Right, Arithmetic Shift Right and Rotate Right), but Logical Shift Left is the most useful as it allows an array indexed to be scaled by size of each array element.

As well as the three forms of offset, the offset and base register are used in three different ways to form the memory address.

1    Offset addressing:
The base register and offset are simply added or subtracted to form the memory address.

2    Pre-indexed addressing:
The base register and offset are added or subtracted to form the memory address. The base register is then updated with this new address, to allow automatic indexing through an array or memory block.

3    Post-indexed addressing:
The value of the base register alone is used as the memory address. The base register and offset are added or subtracted and this value is stored back in the base register, to allow automatic indexing through an array or memory block.

# The ARM Instruction Set

## 3.8.1 Examples

```
LDR    R1, [R0]              ; Load register 1 from the address in register 0
LDR    R8, [R3, #4]          ; Load R8 from the address in R3 + 4
LDR    R12, [R13, #-4]       ; Load R12 from R13 - 4
STR    R2, [R1, #0x100]      ; Store R2 to the address in R1 + 0x100

LDRB   R5, [R9]              ; Load a byte into R5 from R9 (zero top 3 bytes)
LDRB   R3, [R8, #3]          ; Load byte to R3 from R8 + 3 (zero top 3 bytes)
STRB   R4, [R10, #0x200]     ; Store byte from R4 to R10 + 0x200

LDR    R11, [R1, R2]         ; Load R11 from the address in R1 + R2
STRB   R10, [R7, -R4]        ; Store byte from R10 to the address in R7 - R4

LDR    R11,[R3,R5,LSL #2]    ; Load R11 from R3 + (R5 x 4)
LDR    R1, [R0, #4]!         ; Load R1 from R0 + 4, then R0 = R0 + 4
STRB   R7, [R6, #-1]!        ; Store byte from R7 to R6 - 1, then R6 = R6 - 1

LDR    R3, [R9], #4          ; Load R3 from R9, then R9 = R9 + 4
STR    R2, [R5], #8          ; Store word from R2 to R5, then R5 = R5 + 8

LDR    R0, [PC, #40]         ; Load R0 from PC + 8 + 0x40
LDR    R0, [R1], R2          : Load R0 from R1, then R1 = R1 + R2
```

## 3.8.2 Examples of halfword and signed byte addressing modes

The Halfword and Signed Byte addressing modes are a subset of the above addressing modes. The scaled register offset is not supported, and the immediate offset contains 8 bits, not 12.

```
LDRH   R1, [R0]              ; Load a halfword to R1 from R0 (zero top bytes)
LDRH   R8, [R3, #2]          ; Load a halfword into R8 from R3 + 2
LDRH   R12, [R13, #-6]       ; Load a halfword R12 from R13 - 6
STRH   R2, [R1, #0x80]       ; Store halfword from R2 to R1 + 0x80

LDRSH  R5, [R9]              ; Load signed halfword to R5 from R9
LDRSB  R3, [R8, #3]          ; Load signed byte to R3 from R8 + 3
LDRSB  R4, [R10, #0xc1]      ; Load signed byte to R4 from R10 + 0xc1

LDRH   R11, [R1, R2]         ; Load halfword R11 from the address in R1 + R2
STRH   R10, [R7, -R4]        ; Store halfword from R10 to R7 - R4

LDRSH  R1, [R0, #2]!         ; Load signed halfword R1 from R0+2, then R0=R0+2
LDRSB  R7, [R6, #-1]!        ; Load signed byte to R7 from R6-1, then R6=R6-1

LDRH   R3, [R9], #2          ; Load halfword to R3 from R9, then R9 = R9 + 2
STRH   R2, [R5], #8          ; Store halfword from R2 to R5, then R5 = R5 + 8
```

**ARM Architecture Reference Manual**
ARM DUI 0100B

## 3.9    Load and Store Word or Unsigned Byte Instructions

Load instructions load a single value from memory and write it to a general-purpose register.

Store instructions read a value from a general-purpose register and store it to memory.

Load and store instructions have a single instruction format:

```
LDR|STR{<cond>}{B}  Rd, <addressing_mode>
```

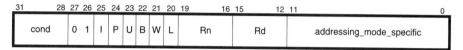

**The I, P, U and W bits:** These bits distinguish between different types of `<addressing_mode>`.

**The L bit:** This bit distinguishes between a Load (L==1) and a Store instruction (L==0).

### 3.9.1    List of load and store word or unsigned byte instructions

| | | |
|---|---|---|
| LDR | Load word | page 3-44 |
| LDRB | Load byte | page 3-45 |
| LDRBT | Load byte with user mode privilege | page 3-46 |
| LDRT | Load word with user mode privilege | page 3-50 |
| STR | Store word | page 3-69 |
| STRB | Store byte | page 3-70 |
| STRBT | Store byte with user mode privilege | page 3-71 |
| STRT | Store word with user mode privilege | page 3-73 |

# The ARM Instruction Set

## 3.10 Load and Store Halfword and Load Signed Byte Instructions

Load instructions load a single value from memory and write it to a general-purpose register.

Store instructions read a value from a general-purpose register and store it to memory.

Load and store halfword and load signed byte instructions have a single instruction format:

```
LDR|STR{<cond>}H|SH|SB  Rd, <addressing_mode>
```

| 31 | 28 | 27 26 25 | 24 | 23 | 22 | 21 | 20 | 19 | 16 | 15 | 12 | 11 | 8 | 7 | 6 | 5 | 4 | 3 | 0 |
|---|---|---|---|---|---|---|---|---|---|---|---|---|---|---|---|---|---|---|---|
| cond | | 0  0  0 | P | U | 1 | W | L | Rn | | Rd | | addr_mode | | 1 | S | H | 1 | addr_mode | |

**The addr_mode bits:** These bits are addressing mode specific.

**The I, P, U and W bits:** These bits specify the type of `<addressing_mode>` (see section *3.18 Load and Store Halfword or Load Signed Byte Addressing Modes* on page 3-109).

**The L bit:** This bit distinguishes between a Load (L==1) and a Store instruction (L==0).

**The S bit:** This bit distinguishes between a signed (S==1) and an unsigned (S==0) halfword access.

**The H bit:** This bit distinguishes between a halfword (H==1) and a signed byte (H==0) access.

## 3.10.1 List of load and store halfword and load signed byte instructions

**ARM Architecture Reference Manual**
ARM DUI 0100B

## 3.11 Load and Store Multiple Instructions

Load Multiple instructions load a subset (or possibly all) of the general-purpose registers from memory.

Store Multiple instructions store a subset (or possibly all) of the general-purpose registers to memory.

Load and Store Multiple instructions have a single instruction format.

```
LDM{<cond>}<addressing_mode>  Rn{!}, <register_list>{^}
STM{<cond>}<addressing_mode>  Rn{!}, <registers>{^}
```

where:

```
<addressing_mode> = IA | IB | DA | DB | FD | FA | ED | EA
```

| 31 | 28 | 27 | 26 | 25 | 24 | 23 | 22 | 21 | 20 | 19 | 16 | 15 | 0 |
|---|---|---|---|---|---|---|---|---|---|---|---|---|---|
| cond | | 1 | 0 | 0 | P | U | S | W | L | Rn | | register list | |

**The register list:** The `<register_list>` has one bit for each general-purpose register; bit 0 is for register zero, and bit 15 is for register 15 (the PC).

The register syntax list is an opening bracket, followed by a comma-separated list of registers, followed by a closing bracket. A sequence of consecutive registers may be specified by separating the first and last registers in the range with a minus sign.

**The P, U and W bits:** These bits distinguish between the different types of addressing mode. See *3.19 Load and Store Multiple Addressing Modes* on page 3-116.

**The S bit:** For LDMs that load the PC, the S bit indicates that the CPSR is loaded from the SPSR. For LDMs that do not load the PC and all STMs, it indicates that when the processor is in a privileged mode, the user-mode banked registers are transferred and not the registers of the current mode.

**The L bit:** This bit distinguishes between a Load (L==1) and a Store (L==0) instruction.

### Addressing modes

For full details of the addressing modes for these instructions, refer to *3.19 Load and Store Multiple Addressing Modes* on page 3-116, and *3.20 Load and Store Multiple Addressing Modes (Alternative names)* on page 3-121.

### 3.11.1 Examples

```
STMFD    R13!,    {R0 - R12, LR}
LDMFD    R13!,    {R0 - R12, PC}
LDMIA    R0,      {R5 - R8}
STMDA    R1!,     {R2, R5, R7 - R9, R11}
```

# The ARM Instruction Set

## 3.11.2   List of load and store multiple instructions

**ARM Architecture Reference Manual**
ARM DUI 0100B

## 3.12 Semaphore Instructions

The ARM instruction set has two semaphore instructions:

- Swap (SWP)
- Swap Byte (SWPB)

These instructions are provided for process synchronisation. Both instructions generate an atomic load and store operation, allowing a memory semaphore to be loaded and altered without interruption.

SWP and SWPB have a single addressing mode; the address is the contents of a register. Separate registers are used to specify the value to store and the destination of the load; if the same register is specified SWP exchanges the value in the register and the value in memory.

The semaphore instructions do not provide a compare and conditional write facility; this must be done explicitly.

### Examples

```
SWP     R12, R10, [R9]    ; load R12 from address R9 and
                          ; store R10 to address R9

SWPB    R3, R4, [R8]      ; load byte to R3 from address R8 and
                          ; store byte from R4 to address R8

SWP     R1, R1, [R2]      ; Exchange value in R1 and address in R2
```

### 3.12.1 List of semaphore instructions

# The ARM Instruction Set

## 3.13 Coprocessor Instructions

**Note:** *Coprocessor instructions are not implemented in Architecture version 1.*

The ARM instruction set provides 3 types of instruction for communicating with coprocessors. The instruction set distinguishes up to 16 coprocessors with a 4-bit field in each coprocessor instruction, so each coprocessor is assigned a particular number (one coprocessor can use more than one of the 16 numbers if a large coprocessor instruction set is required).

The three classes of coprocessor instruction allow:

- ARM to initiate a coprocessor data processing operation
- ARM registers to be transferred to and from coprocessor registers
- ARM to generate addresses for the coprocessor load and store instructions

Coprocessors execute the same instruction stream as ARM, ignoring ARM instructions and coprocessor instructions for other coprocessors. Coprocessor instructions that cannot be executed by coprocessor hardware cause an UNDEFINED instruction trap, allowing software emulation of coprocessor hardware.

A coprocessor can partially execute an instruction and then cause an exception; this is useful for handling run-time-generated exceptions (like divide-by-zero or overflow).

Not all fields in coprocessor instructions are used by ARM; coprocessor register specifiers and opcodes are defined by individual coprocessors. Therefore, only generic instruction mnemonics are provided for coprocessor instructions; assembler macros can be used to transform custom coprocessor mnemonics into these generic mnemonics (or to regenerate the opcodes manually).

### Examples

```
CDP     p5, 2, c12, c10, c3, 4 ; Coprocessor 5 data operation
                               ; opcode 1 = 2, opcode 2 = 4
                               ; destination register is 12
                               ; source registers are 10 and 3

MRC     p15, 5, R4, c0, c2, 3  ; Coprocessor 15 transfer to ARM register
                               ; opcode 1 = 5, opcode 2 = 3
                               ; ARM destination register = R4
                               ; coproc source registers are 0 and 2

MCR     p14, 1, R7, c7, c12, 6 ; ARM register transfer to Coprocessor 14
                               ; opcode 1 = 1, opcode 2 = 6
                               ; ARM source register = R7
                               ; coproc dest registers are 7 and 12

LDC     p6, CR1, [R4]          ; Load from memory to coprocessor 6
                               ; ARM register 4 contains the address
                               ; Load to CP reg 1

LDC     p6, CR4, [R2, #4]      ; Load from memory to coprocessor 6
                               ; ARM register R2 + 4 is the address
                               ; Load to CP reg 4
```

**ARM Architecture Reference Manual**
ARM DUI 0100B

```
STC     p8, CR8, [R2, #4]!      ; Store from coprocessor 8 to memory
                                ; ARM register R2 + 4 is the address
                                ; after the transfer R2 = R2 + 4
                                ; Store from CP reg 8

STC     p8, CR9, [R2], #-16     ; Store from coprocessor 8 to memory
                                ; ARM register R2 holds the address
                                ; after the transfer R2 = R2 - 16
                                ; Store from CP reg 9
```

## 3.13.1    List of coprocessor instructions

# The ARM Instruction Set

## 3.14 Extending the Instruction Set

The ARM instruction set can be extended in four areas:

- Arithmetic instruction extension space
- Control instruction extension space
- Load/store instruction extension space
- Coprocessor instruction extension space

Currently, these instructions are UNDEFINED (they cause an undefined instruction exception). These parts of the address space will be used in the future to add new instructions.

### 3.14.1 Arithmetic instruction extension space

Instructions with the following opcodes are the arithmetic instruction extension space:

```
opcode[27:24]   = 0
opcode[7:4]     = 0b1001
```

The field names given are only guidelines, which are likely to simplify implementation.

| 31 | 28 | 27 26 25 24 | 23 | 20 19 | 16 15 | 12 11 | 8 7 6 5 4 | 3 | 0 |
|----|----|----|----|----|----|----|----|----|----|
| cond | | 0 0 0 0 | op1 | Rn | Rd | Rs | 1 0 0 1 | Rm | |

#### MUL and MLA

Multiply and Multiply Accumulate (MUL and MLA) instructions use op1 values 0 to 3.

| Rn | specifies the destination register |
|----|----|
| Rm and Rs | specify source registers |
| Rd | specifies the accumulated value |

#### UMULL, UMLAL, SMULL, SMLAL

The Signed and Unsigned Multiple Long and Multiply Accumulate Long (UMULL, UMLAL, SMULL, SMLAL) instructions use op1 values 8 to 15;

| Rn and Rd | specify the destination registers |
|----|----|
| Rm and Rs | specify the source registers |

#### Other opcodes

The meaning of all other opcodes in the arithmetic instruction extension space is:

- UNDEFINED on ARM Architecture 4
- UNPREDICTABLE on earlier versions of the architecture.

**ARM Architecture Reference Manual**
ARM DUI 0100B

## 3.14.2   Control instruction extension space

Instructions with:

    opcode[27:26]   =   0b00

    opcode[24:23]   =   0b10

    opcode[20]      =   0

and not:

    opcode[25]      =   0

    opcode[7]       =   1

    opcode[4]       =   1

are the control instruction extension space. The field names given are only guidelines, which are likely to simplify implementation.

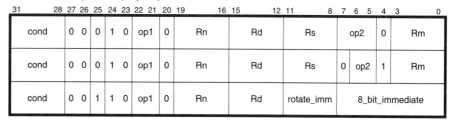

### MRS

The Move status register to general-purpose register (MRS) instruction sets:

    opcode[25]      =   0

    opcode[19:16]   =   0b1111

    opcode[11:0]    =   0

and uses both op1 = 0b00 and op1 = 0b10 .
Rd is used to specify the destination register.

### MSR

The Move general-purpose register to status register (MSR) instruction sets:

    opcode[25]      =   0

    opcode[15:12]   =   0b1111

    opcode[11:4]    =   0

and uses both op1 = 0b01 and op1= 0b11.

    Rm                      specifies the source register
    opcode[19:16]       ·   hold the write mask

# The ARM Instruction Set

The Move immediate value to status register (MSR) instruction sets:

```
opcode[25]      =  1

opcode[15:12]   =  0b1111
```

and uses both `op1 = 0b01` and `op1= 0b11`.

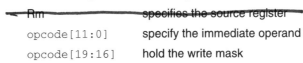

```
opcode[11:0]      specify the immediate operand

opcode[19:16]     hold the write mask
```

### BX

The Branch and Exchange Instruction Set (BX) instruction sets:

```
opcode[25]      =  0

opcode[19:8]    =  0b111111111111

opcode[7:4]     =  0b0001
```

and uses `op1 = 0b01`.

Rm is used to specify the source register.

### Other opcodes

The meaning of all other opcodes in the control instruction extension space is:

- UNDEFINED on ARM Architecture 4
- UNPREDICTABLE on earlier versions of the architecture

## 3.14.3 Load/Store instruction extension space

Instructions with

```
opcode[27:25]   =  0b000

opcode[7]       =  1

opcode[4]       =  1
```

and not

```
opcode[24]      =  0

opcode[6:5]     =  0
```

are the load/store instruction extension space.

The field names given are only guidelines, which are likely to simplify implementation.

| 31 | 28 | 27 26 25 | 24 | 23 | 22 | 21 | 20 | 19　　16 | 15　　12 | 11　　8 | 7 | 6 5 4 | 3　　0 |
|---|---|---|---|---|---|---|---|---|---|---|---|---|---|
| cond | | 0　0　0 | P | U | B | W | L | Rn | Rd | Rs | 1 | op1　1 | Rm |

**ARM Architecture Reference Manual**
ARM DUI 0100B

**SWP and SWPB**

The Swap and Swap Byte (SWP and SWPB) instructions set:

```
opcode[24:23]   =   0b10
opcode[21:20]   =   0b00
opcode[11:4]    =   0b00001001
```

where:

| | |
|---|---|
| Rn | specifies the base address |
| Rd | specifies the destination register |
| Rm | specifies the source register |
| Opcode[22] | indicates a byte transfer |

**LDRH**

The Load Halfword (LDRH) instruction sets:

```
opcode[20]      =   1
op1             =   0b01
```

where:

| | |
|---|---|
| the B bit | distinguishes between a register and an immediate offset |
| the P, U and W bits | specify the addressing mode |
| Rn | specifies the base register |
| Rd | specifies the destination register |
| Rm | specifies a register offset |
| Rs and Rm | specify an immediate offset |

**LDRSH**

The Load Signed Halfword (LDRSH) instruction sets:

```
opcode[20]      =   1
op1             =   0b11
```

where:

| | |
|---|---|
| the B bit | distinguishes between a register and an immediate offset |
| the P, U and W bits | specify the addressing mode |
| Rn | specifies the base register |
| Rd | specifies the destination register |
| Rm | specifies a register offset |
| Rs and Rm | specify an immediate offset |

# The ARM Instruction Set

**LDRSB**

The Load Signed Byte (LDRSB) instruction sets:

```
opcode[20]    = 1
op1           = 0b10
```

where:

| | |
|---|---|
| the B bit | distinguishes between a register and an immediate offset |
| the P, U and W bits | specify the addressing mode |
| Rn | specifies the base register |
| Rd | specifies the destination register |
| Rm | specifies a register offset |
| Rs and Rm | specify an immediate offset |

**STRH**

The Store Halfword (STRH) instruction sets:

```
opcode[20]    = 0
op1           = 0b01
```

where:

| | |
|---|---|
| the B bit | distinguishes between a register and an immediate offset |
| the P, U and W bits | specify the addressing mode |
| Rn | specifies the base register |
| Rd | specifies the source register |
| Rm | specifies a register offset |
| Rs and Rm | specify an immediate offset |

**Other opcodes**

The meaning of all other opcodes in the Load/Store instruction extension space is:

- UNDEFINED on ARM Architecture 4
- UNPREDICTABLE on earlier versions of the architecture

**ARM Architecture Reference Manual**
ARM DUI 0100B

**3.14.4    Coprocessor instruction extension space**

Instructions with

opcode[27:24]   =   0b1100

opcode[21]      =   0

are the coprocessor instruction extension space. The names given to fields are only guidelines, which if followed are likely to simplify implementation.

| 31 | 28 | 27 26 25 24 | 23 | 22 | 21 | 20 | 19 | 16 | 15 | 12 | 11 | 8 | 7 | 0 |
|---|---|---|---|---|---|---|---|---|---|---|---|---|---|---|
| cond | | 1  1  0  0 | X | X | 0 | X | Rn | | CRd | | cp_num | | offset | |

The meaning of instructions in the coprocessor instruction extension space is:

- UNDEFINED on ARM Architecture 4
- UNPREDICTABLE on earlier versions of the architecture

**3.14.5    Undefined instruction Space**

Instructions with

opcode[27:25]   =   0b011

opcode[4]       =   1

are UNDEFINED instruction space.

| 31 | 28 | 27 26 25 | 24 | 5 | 4 | 3 | 0 |
|---|---|---|---|---|---|---|---|
| cond | | 0  1  1 | x x x x  x x x x  x x x x  x x x x  x x x x  x x x | 1 | x  x  x  x | |

The meaning of instructions in the UNDEFINED instruction space is UNDEFINED on all versions of the ARM Architecture.

# ARM Instructions

Instruction name
given in the following alphabetical list

Description

Syntax

Functional area
described in the preceding section of this chapter

Addressing mode
indicates if an addressing mode applies to this instruction

Architecture availability
indicates if there is a restriction on availability
Not all instructions are available in
all versions of the ARM architecture

Encoding
specifies the bit patterns for the instruction

Operation
describes the operation of the instruction in pseudo-code

Exceptions
lists any possible exceptions

Qualifiers and flag settings
lists any conditions and flag settings
that apply to the instruction

User notes
gives notes on using the instruction

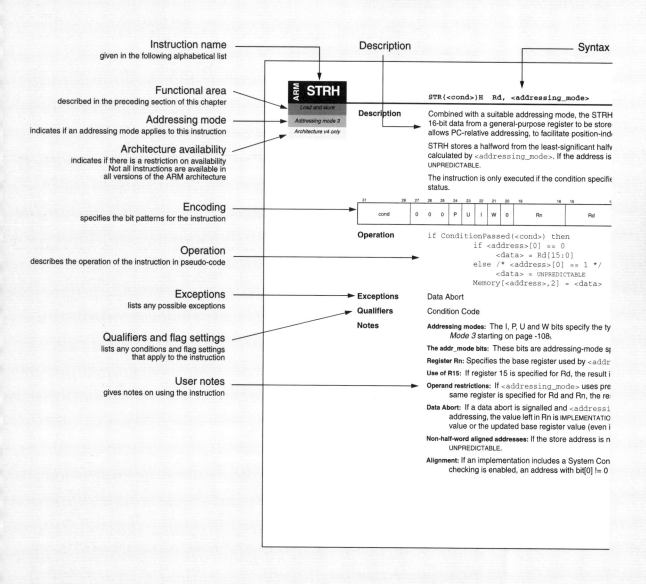

**ARM STRH**

Load and store
Addressing mode 3
Architecture v4 only

`STR{<cond>}H  Rd, <addressing_mode>`

**Description**

Combined with a suitable addressing mode, the STRH
16-bit data from a general-purpose register to be store
allows PC-relative addressing, to facilitate position-ind

STRH stores a halfword from the least-significant halfv
calculated by `<addressing_mode>`. If the address is
UNPREDICTABLE.

The instruction is only executed if the condition specifie
status.

| 31 | 28 | 27 | 26 | 25 | 24 | 23 | 22 | 21 | 20 | 19 | 16 | 15 | 1 |
|---|---|---|---|---|---|---|---|---|---|---|---|---|---|
| cond | | 0 | 0 | 0 | P | U | I | W | 0 | Rn | | Rd | |

**Operation**

```
if ConditionPassed(<cond>) then
        if <address>[0] == 0
            <data> = Rd[15:0]
        else /* <address>[0] == 1 */
            <data> = UNPREDICTABLE
        Memory[<address>,2] = <data>
```

**Exceptions**   Data Abort

**Qualifiers**   Condition Code

**Notes**

**Addressing modes:** The I, P, U and W bits specify the ty
*Mode 3* starting on page -108).

**The addr_mode bits:** These bits are addressing-mode sp

**Register Rn:** Specifies the base register used by `<addr`

**Use of R15:** If register 15 is specified for Rd, the result i

**Operand restrictions:** If `<addressing_mode>` uses pre
same register is specified for Rd and Rn, the res

**Data Abort:** If a data abort is signalled and `<addressi`
addressing, the value left in Rn is IMPLEMENTATIO
value or the updated base register value (even i

**Non-half-word aligned addresses:** If the store address is n
UNPREDICTABLE.

**Alignment:** If an implementation includes a System Con
checking is enabled, an address with bit[0] != 0

`ADC{<cond>}{S}  Rd, Rn, <shifter_operand>`

**Description**

The ADC (Add with Carry) instruction adds the value of `<shifter_operand>` and the value of the Carry flag to the value of register Rn, and stores the result in the destination register Rd. The condition code flags are optionally updated (based on the result).

ADC is used to synthesize multi-word addition. If register pairs R0,R1 and R2,R3 hold 64-bit values (where 0 and R2 hold the least-significant words), the following instructions leave the 64-bit sum in R4,R5:

```
ADDS    R4,R0,R2
ADC     R5,R1,R3
```

The instruction:

```
ADCS R0,R0,R0
```

produces a single-bit Rotate Left with Extend operation (33-bit rotate though the carry flag) on R0. See 3-97 for more information.

The instruction is only executed if the condition specified in the instruction matches the condition code status. The conditions are defined in *3.3 The Condition Field* on page 3-4.

| 31      28 | 27 26 | 25 | 24 23 22 21 | 20 | 19      16 | 15      12 | 11      0 |
|------------|-------|-----|-------------|-----|------------|------------|-----------|
| cond | 0  0 | I | 0  1  0  1 | S | Rn | Rd | shifter_operand |

**Operation**

```
if ConditionPassed(<cond>) then
    Rd = Rn + <shifter_operand> + C Flag
    if S == 1 and Rd == R15 then
        CPSR = SPSR
    else if S == 1 then
        N Flag = Rd[31]
        Z Flag = if Rd == 0 then 1 else 0
        C Flag = CarryFrom(Rn + <shifter_operand> + C Flag)
        V Flag = OverflowFrom (Rn + <shifter_operand> + C Flag)
```

**Exceptions**    None

**Qualifiers**    Condition Code

S updates condition code flags N,Z,C,V

**Notes**

**Shifter operand:** The shifter operands for this instruction are given in *Addressing Mode 1* starting on page 3-84.

**The I bit:** Bit 25 is used to distinguish between the immediate and register forms of `<shifter_operand>`.

**Writing to R15:** When Rd is R15 and the S flag in the instruction is not set, the result of the operation is placed in the PC. When Rd is R15 and the S flag is set, the result of the operation is placed in the PC and the SPSR corresponding to the current mode is moved to the CPSR. This allows state changes which atomically restore both PC and CPSR. This form of the instruction is UNPREDICTABLE in User mode and System mode.

**ARM Architecture Reference Manual**

ARM DUI 0100B

```
ADD{<cond>}{S}  Rd, Rn, <shifter_operand>
```

**Description**   The ADD instruction adds the value of `<shifter_operand>` to the value of register Rn, and stores the result in the destination register Rd. The condition code flags are optionally updated (based on the result).

ADD is used to add two values together to produce a third.

To increment a register value (in Rx), use:

```
ADD Rx, Rx, #1
```

Constant multiplication (of Rx) by $2^n+1$ (into Rd) can be performed with:

```
ADD Rd, Rx, Rx LSL #n
```

To form a PC-relative address, use:

```
ADD Rs, PC, #offset
```

The instruction is only executed if the condition specified in the instruction matches the condition code status. The conditions are defined in *3.3 The Condition Field* on page 3-4.

| 31    28 | 27 26 | 25 | 24 23 22 21 | 20 | 19    16 | 15    12 | 11                    0 |
|----------|-------|----|----|----|--------|--------|------------------------|
| cond | 0 0 | I | 0 1 0 0 | S | Rn | Rd | shifter_operand |

**Operation**

```
if ConditionPassed(<cond>) then
        Rd = Rn + <shifter_operand>
        if S == 1 and Rd == R15 then
            CPSR = SPSR
        else if S == 1 then
            N Flag = Rd[31]
            Z Flag = if Rd == 0 then 1 else 0
            C Flag = CarryFrom(Rn + <shifter_operand>)
            V Flag = OverflowFrom (Rn + <shifter_operand>)
```

**Exceptions**   None

**Qualifiers**   Condition Code

S updates condition code flags N,Z,C,V

**Notes**   **Shifter operand:** The shifter operands for this instruction are given in *Addressing Mode 1* starting on page 3-84.

**The I bit:** Bit 25 is used to distinguish between the immediate and register forms of `<shifter_operand>`.

**Writing to R15:** When Rd is R15 and the S flag in the instruction is not set, the result of the operation is placed in the PC. When Rd is R15 and the S flag is set, the result of the operation is placed in the PC and the SPSR corresponding to the current mode is moved to the CPSR. This allows state changes which atomically restore both PC and CPSR. This form of the instruction is UNPREDICTABLE in User mode and System mode.

# AND

AND{<cond>}{S}  Rd, Rn, <shifter_operand>

*Data processing*

*Addressing mode 1*

**Description**

The AND instruction performs a bitwise AND of the value of register Rn with the value of <shifter_operand>, and stores the result in the destination register Rd. The condition code flags are optionally updated (based on the result).

AND is most useful for extracting a field from a register, by ANDing the register with a mask value that has 1's in the field to be extracted, and 0's elsewhere.

The instruction is only executed if the condition specified in the instruction matches the condition code status. The conditions are defined in *3.3 The Condition Field* on page 3-4.

| 31      28 | 27 | 26 | 25 | 24 | 23 | 22 | 21 | 20 | 19        16 | 15        12 | 11                    0 |
|------------|----|----|----|----|----|----|----|----|--------------|--------------|-------------------------|
| cond       | 0  | 0  | I  | 0  | 0  | 0  | 0  | S  | Rn           | Rd           | shifter_operand         |

**Operation**

```
if ConditionPassed(<cond>) then
    Rd = Rn AND <shifter_operand>
    if S == 1 and Rd == R15 then
        CPSR = SPSR
    else if S == 1 then
        N Flag = Rd[31]
        Z Flag = if Rd == 0 then 1 else 0
        C Flag = <shifter_carry_out>
        V Flag = unaffected
```

**Exceptions**

None

**Qualifiers**

Condition Code

S updates condition code flags N,Z,C

**Notes**

**Shifter operand:** The shifter operands for this instruction are given in *Addressing Mode 1* starting on page 3-84.

**The I bit:** Bit 25 is used to distinguish between the immediate and register forms of <shifter_operand>.

**Writing to R15:** When Rd is R15 and the S flag in the instruction is not set, the result of the operation is placed in the PC. When Rd is R15 and the S flag is set, the result of the operation is placed in the PC and the SPSR corresponding to the current mode is moved to the CPSR. This allows state changes which atomically restore both PC and CPSR. This form of the instruction is UNPREDICTABLE in User mode and System mode.

**ARM Architecture Reference Manual**

ARM DUI 0100B

**Description**  The B (Branch) and BL (Branch and Link) instructions provide both conditional and unconditional changes to program flow. The Branch with Link instruction is used to perform a subroutine call; the return from subroutine is achieved by copying the LR to the PC.

B and BL cause a branch to a target address. The branch target address is calculated by:

1    shifting the 24-bit signed (two's complement) offset left two bits

2    sign-extending the result to 32 bits

3    adding this to the contents of the PC (which contains the address of the branch instruction plus 8)

The instruction can therefore specify a branch of +/- 32Mbytes.

In the BL variant of the instruction, the L (link) bit is set, and the address of the instruction following the branch is copied into the link register (R14).

The instruction is only executed if the condition specified in the instruction matches the condition code status. The conditions are defined in *3.3 The Condition Field* on page 3-4.

| 31      28 | 27 26 25 | 24 | 23                                    0 |
|------------|----------|----|-----------------------------------------|
| cond       | 1  0  1  | L  | 24_bit_signed_offset                    |

**Operation**

```
if ConditionPassed(<cond>) then
    if L == 1 then
        LR = address of the instruction after the branch instruction
    PC = PC + (SignExtend(<24_bit_signed_offset>) << 2)
```

**Exceptions**    None

**Qualifiers**    Condition Code

L (Link) stores a return address in the LR (R14) register

**Notes**    **Offset calculation:** An assembler will calculate the branch offset address from the difference between the address of the current instruction and the address of the target (given as a program label) minus eight (because the PC holds the address of the current instruction plus eight).

**Memory bounds:** Branching backwards past location zero and forwards over the end of the 32-bit address space is UNPREDICTABLE.

**ARM Architecture Reference Manual**
ARM DUI 0100B

BIC{<cond>}{S}  Rd, Rn, <shifter_operand>

**Description**    The BIC (Bit Clear) instruction performs a bitwise AND of the value of register Rn with the complement of the value of <shifter_operand>, and stores the result in the destination register Rd. The condition code flags are optionally updated (based on the result).

BIC can be used to clear selected bits in a register; for each bit, BIC with 1 will clear the bit, BIC with 0 will leave it unchanged.

The instruction is only executed if the condition specified in the instruction matches the condition code status. The conditions are defined in *3.3 The Condition Field* on page 3-4.

| 31      28 | 27 | 26 | 25 | 24 | 23 | 22 | 21 | 20 | 19      16 | 15      12 | 11                0 |
|------------|----|----|----|----|----|----|----|----|------------|------------|---------------------|
| cond | 0 | 0 | I | 1 | 1 | 1 | 0 | S | Rn | Rd | shifter_operand |

**Operation**
```
if ConditionPassed(<cond>) then
    Rd = Rn AND NOT <shifter_operand>
    if S == 1 and Rd == R15 then
        CPSR = SPSR
    else if S == 1 then
        N Flag = Rd[31]
        Z Flag = if Rd == 0 then 1 else 0
        C Flag = <shifter_carry_out>
        V Flag = unaffected
```

**Exceptions**    None

**Qualifiers**    Condition Code

S updates condition code flags N,Z,C

**Notes**    **Shifter operand:** The shifter operands for this instruction are given in *Addressing Mode 1* starting on page 3-84.

**Writing to R15:**  When Rd is R15 and the S flag in the instruction is not set, the result of the operation is placed in the PC. When Rd is R15 and the S flag is set, the result of the operation is placed in the PC and the SPSR corresponding to the current mode is moved to the CPSR. This allows state changes which atomically restore both PC and CPSR. This form of the instruction is UNPREDICTABLE in User mode and System mode.

**The I bit:**  Bit 25 is used to distinguish between the immediate and register forms of <shifter_operand>.

**ARM Architecture Reference Manual**

ARM DUI 0100B

**Description**

The BX (Branch and Exchange instructions set) is UNDEFINED on ARM Architecture Version 4. On ARM Architecture Version 4T, this instruction branches and selects the instruction set decoder to use to decode the instructions at the branch destination. The branch target address is the value of register Rm. The T flag is updated with bit 0 of the value of register Rm.

BX is used to branch between ARM code and THUMB code. On ARM Architecture 4, it causes an UNDEFINED instruction exception to allow the THUMB instruction set to be emulated.

The instruction is only executed if the condition specified in the instruction matches the condition code status. The conditions are defined in *3.3 The Condition Field* on page 3-4.

| 31    28 | 27 26 25 24 23 22 21 20 | 19    16 | 15    12 | 11    8 | 7 6 5 4 | 3    0 |
|----------|--------------------------|----------|----------|---------|---------|--------|
| cond | 0 0 0 1 0 0 1 0 | SBO | SBO | SBO | 0 0 0 1 | Rm |

**Operation**

```
if ConditionPassed(<cond>) then
        T Flag = Rm[0]
        PC = Rm[31:1] << 1
```

**Exceptions**       None

**Operation**       Condition Code

**Notes**       **Transferring to THUMB:** When transferring to the THUMB instruction set, bit[0] of PC will be cleared (set to zero), and bits[31:1] will be copied from Rm to the PC.

**Transferring to ARM:** When transferring to the ARM instruction set, bit[0] of PC will be cleared (set to zero), and bits[31:1] will be copied from Rm to the PC. If bit[1] of Rm is set, the result is UNPREDICTABLE.

# CDP

CDP{<cond>}  p<cp#>, <opcode_1>, CRd, CRn, CRm, <opcode_2>

*Coprocessor*

*Not in architecture v1*

**Description**   The CDP (Coprocessor Data Processing) instruction tells the coprocessor specified by <cp#> to perform an operation that is independent of ARM registers and memory. If no coprocessors indicate that they can execute the instruction, an UNDEFINED instruction exception is generated.

CDP is used to initiate coprocessor instructions that do not operate on values in ARM registers or in main memory; for example, a floating-point multiply instruction for a floating-point accelerator coprocessor.

The instruction is only executed if the condition specified in the instruction matches the condition code status. The conditions are defined in *3.3 The Condition Field* on page 3-4.

| 31      28 | 27 26 25 24 | 23 22 21 20 | 19      16 | 15      12 | 11      8 | 7      5 | 4 | 3      0 |
|------------|-------------|-------------|------------|------------|-----------|----------|---|----------|
| cond | 1 1 1 0 | opcode_1 | CRn | CRd | cp_num | opcode_2 | 0 | CRm |

**Operation**   `if ConditionPassed(<cond>) then`
`        Coprocessor[<cp_num>] dependent operation`

**Exceptions**   Undefined Instruction

**Qualifiers**   Condition Code

**Notes**   **Coprocessor fields:** Only instruction bits[31:24], bits[11:8] and bit[4] are which architecture defined; the remaining fields are only recommendations, which if followed, will be compatible with ARM Development Systems.

ARM

CMN{<cond>}   Rn, <shifter_operand>

**Description**   The CMN (Compare Negative) instruction compares an arithmetic value and the negative of an arithmetic value (an immediate or the value of a register) and sets the condition code flags so that subsequent instructions can be conditionally executed.

CMN performs a comparison by adding (or subtracting the negative of) the value of <shifter_operand> to (from) the value of register Rn, and updates the condition code flags (based on the result). The comparison is the subtraction of the negative of the second operand from the first operand (which is the same as adding the two operands).

The instruction is only executed if the condition specified in the instruction matches the condition code status. The conditions are defined in *3.3 The Condition Field* on page 3-4.

| 31 28 | 27 26 | 25 | 24 23 22 21 | 20 | 19 16 | 15 12 | 11 0 |
|---|---|---|---|---|---|---|---|
| cond | 0 0 | I | 1 0 1 1 | 1 | Rn | SBZ | shifter_operand |

**Operation**
```
if ConditionPassed(<cond>) then
        <alu_out> = Rn + <shifter_operand>
        N Flag = <alu_out>[31]
        Z Flag = if <alu_out> == 0 then 1 else 0
        C Flag = CarryFrom(Rn + <shifter_operand>)
        V Flag = OverflowFrom (Rn + <shifter_operand>)
```

**Exceptions**   None

**Qualifiers**   Condition Code

**Notes**   **Shifter operand:** The shifter operands for this instruction are given in *Addressing Mode 1* starting on page 3-84.

**The I bit:** Bit 25 is used to distinguish between the immediate and register forms of <shifter_operand>.

**ARM Architecture Reference Manual**
ARM DUI 0100B

# CMP

**Data processing**

**Addressing mode 1**

CMP{<cond>}  Rn, <shifter_operand>

**Description**

The CMP (Compare) instruction compares two arithmetic values and sets the condition code flags so that subsequent instructions can be conditionally executed. The comparison is a subtraction of the second operand from the first operand.

CMP performs a comparison by subtracting the value of <shifter_operand> from the value of register Rn and updates the condition code flags (based on the result).

The instruction is only executed if the condition specified in the instruction matches the condition code status. The conditions are defined in *3.3 The Condition Field* on page 3-4.

| 31      28 | 27 | 26 | 25 | 24 | 23 | 22 | 21 | 20 | 19      16 | 15      12 | 11      0 |
|---|---|---|---|---|---|---|---|---|---|---|---|
| cond | 0 | 0 | I | 1 | 0 | 1 | 0 | 1 | Rn | SBZ | shifter_operand |

**Operation**

```
if ConditionPassed(<cond>) then
        <alu_out> = Rn - <shifter_operand>
        N Flag = <alu_out>[31]
        Z Flag = if <alu_out> == 0 then 1 else 0
        C Flag = NOT BorrowFrom(Rn - <shifter_operand>)
        V Flag = OverflowFrom (Rn - <shifter_operand>)
```

**Exceptions**    None

**Qualifiers**    Condition Code

**Notes**    **Shifter operand:** The shifter operands for this instruction are given in *Addressing Mode 1* starting on page 3-84.

**The I bit:** Bit 25 is used to distinguish between the immediate and register forms of <shifter_operand>.

**ARM Architecture Reference Manual**

ARM DUI 0100B

| **Description** | The EOR (Exclusive-OR) instruction performs a bitwise Exclusive-OR of the value of register Rn with the value of `<shifter_operand>`, and stores the result in the destination register Rd. The condition code flags are optionally updated (based on the result). |

EOR can be used to invert selected bits in a register; for each bit, EOR with 1 will invert that bit, and EOR with 0 will leave it unchanged.

The instruction is only executed if the condition specified in the instruction matches the condition code status. The conditions are defined in *3.3 The Condition Field* on page 3-4.

| 31 | 28 | 27 | 26 | 25 | 24 | 23 | 22 | 21 | 20 | 19 | 16 | 15 | 12 | 11 | 0 |
|----|----|----|----|----|----|----|----|----|----|----|----|----|----|----|----|
| cond | | 0 | 0 | I | 0 | 0 | 0 | 1 | S | Rn | | Rd | | shifter_operand | |

**Operation**

```
if ConditionPassed(<cond>) then
        Rd = Rn EOR <shifter_operand>
        if S == 1 and Rd == R15 then
            CPSR = SPSR
        else if S == 1 then
            N Flag = Rd[31]
            Z Flag = if Rd == 0 then 1 else 0
            C Flag = <shifter_carry_out>
            V Flag = unaffected
```

**Exceptions**   None

**Qualifiers**   Condition Code
S updates condition code flags N,Z,C

**Notes**   **Shifter operand:** The shifter operands for this instruction are given in *Addressing Mode 1* starting on page 3-84.

**Writing to R15:** When Rd is R15 and the S flag in the instruction is not set, the result of the operation is placed in the PC. When Rd is R15 and the S flag is set, the result of the operation is placed in the PC and the SPSR corresponding to the current mode is moved to the CPSR. This allows state changes which atomically restore both PC and CPSR. This form of the instruction is UNPREDICTABLE in User mode and System mode.

**The I bit:** Bit 25 is used to distinguish between the immediate and register forms of `<shifter_operand>`.

**ARM Architecture Reference Manual**
ARM DUI 0100B

# LDC

*Coprocessor*

*Addressing mode 5*

*Not in Architecture v1*

LDC{<cond>}  p<cp_num>, CRd, <addressing_mode>

**Description** The LDC (Load Coprocessor) instruction is useful to load coprocessor data from memory. The N bit could be used to distinguish between a single- and double-precision transfer for a floating-point load instruction.

LDC loads memory data from the sequence of consecutive memory addresses calculated by <addressing_mode> to the coprocessor specified by <cp_num>. If no coprocessors indicate that they can execute the instruction, an UNDEFINED instruction exception is generated.

The instruction is only executed if the condition specified in the instruction matches the condition code status. The conditions are defined in *3.3 The Condition Field* on page 3-4.

| 31 | 28 | 27 26 25 24 | 23 | 22 | 21 | 20 | 19 | 16 | 15 | 12 | 11 | 8 | 7 | 0 |
|---|---|---|---|---|---|---|---|---|---|---|---|---|---|---|
| cond | | 1 1 0 | P | U | N | W | 1 | Rn | | CRd | | cp_num | | 8_bit_word_offset |

**Operation**
```
if ConditionPassed(<cond>) then
        <address> = <start_address>
        while (NotFinished(coprocessor[<cp_num>]))
            Coprocessor[<cp_num>] = Memory[<address>,4]
            <address> = <address> + 4
        assert <address> == <end_address>
```

**Exceptions** Undefined Instruction; Data Abort

**Qualifiers** Condition Code

**Notes** **Addressing mode:** The P, U and W bits specify the <addressing_mode>. See *Addressing Mode 5* starting on page 3-123.

**The N bit:** This bit is coprocessor-dependent. It can be used to distinguish between two sizes of data to transfer.

**Register Rn:** Specifies the base register used by <addressing_mode>.

**Coprocessor fields:** Only instruction bits[31:23], bits [21:16} and bits[11:0] are ARM architecture-defined; the remaining fields (bit[22] and bits[15:12]) are recommendations for compatibility with ARM Development Systems.

**Data Abort:** If a data abort is signalled and <addressing_mode> uses pre-indexed or post-indexed addressing, the value left in Rn is IMPLEMENTATION DEFINED, but is either the original base register value or the updated base register value.

**Non-word-aligned addresses:** Load coprocessor register instructions ignore the least-significant two bits of <address> (the words are not rotated as for load word).

**Alignment:** If an implementation includes a System Control Coprocessor (see *Chapter 7*), and alignment checking is enabled, an address with bits[1:0] != 0b00 will cause an alignment exception.

**ARM Architecture Reference Manual**

ARM DUI 0100B

**Description**

This form of the LDM (Load Multiple) instruction is useful as a block load instruction (combined with store multiple it allows efficient block copy) and for stack operations, including procedure exit, to restore saved registers, load the PC with the return address, and update the stack pointer.

In this case, LDM loads a non-empty subset (or possibly all) of the general-purpose registers from sequential memory locations. The registers are loaded in sequence, the lowest-numbered register first, from the lowest memory address (<start_addr>); the highest-numbered register last, from the highest memory address (<end_addr>). If the PC is specified in the register list (opcode bit 15 is set), the instruction causes a branch to the address (data) loaded into the PC.

The instruction is only executed if the condition specified in the instruction matches the condition code status. The conditions are defined in *3.3 The Condition Field* on page 3-4.

| 31    28 | 27 26 25 | 24 | 23 | 22 | 21 | 20 | 19    16 | 15                          0 |
|----------|----------|----|----|----|----|----|----------|-------------------------------|
| cond     | 1  0  0  | P  | U  | 0  | W  | 1  | Rn       | register list                 |

**Operation**

```
if ConditionPassed(<cond>) then
        <address> = <start_addr>
        for i = 0 to 15
            if <register_list>[i] == 1
                    Ri = Memory[<address>,4]
                    <address> = <address> + 4
        assert <end_add> == <address> - 4
```

**Exceptions**     Data Abort

**Qualifiers**     Condition Code

! sets the W bit, causing base register update

**Notes**

**Addressing mode:** The P, U and W bits distinguish between the different types of addressing mode. See *Addressing Mode 4* starting on page 3-116.

**Register Rn:** Specifies the base register used by <addressing_mode>.

**Use of R15:** Using R15 as the base register Rn gives an UNPREDICTABLE result.

**Operand restrictions:** If the base register Rn is specified in <register_list>, and writeback is specified, the final value of Rn is UNPREDICTABLE.

**Data Abort:** If a data abort is signalled and <addressing_mode> specifies writeback, the value left in Rn is IMPLEMENTATION DEFINED, but is either the original base register value or the updated base register value (even if Rn is specified in <register_list>). If register 15 is specified in <register_list>, it must not be overwritten if a data abort occurs.

**Non-word-aligned addresses:** Load multiple instructions ignore the least-significant two bits of <address> (the words are not rotated as for load word).

**Alignment:** If an implementation includes a System Control Coprocessor (see *Chapter 7*), and alignment checking is enabled, an address with bits[1:0] != 0b00 will cause an alignment exception.

**ARM Architecture Reference Manual**
ARM DUI 0100B

*Load and store multiple*

*Addressing mode 4*

`LDM{<cond>}<addressing_mode>  Rn, <registers>^`

**Description**

This form of the LDM (Load Multiple) instruction loads user mode registers when the processor is in a privileged mode (useful when performing process swaps).

In this case, LDM instruction loads a non-empty subset (or possibly all except the PC) of the user mode general-purpose registers (which are also the system mode general-purpose registers) from sequential memory locations. The registers are loaded in sequence, the lowest-numbered register first, from the lowest memory address (`<start_address>`); the highest-numbered register last, from the highest memory address (`<end_address>`).

The instruction is only executed if the condition specified in the instruction matches the condition code status. The conditions are defined in *3.3 The Condition Field* on page 3-4.

| 31        28 | 27 26 25 | 24 | 23 | 22 | 21 | 20 19        16 | 15                    0 |
|--------------|----------|----|----|----|----|-----------------|--------------------------|
| cond | 1 0 0 | P | U | 1 | W | 1 | Rn | 0 | register list |

**Operation**

```
if ConditionPassed(<cond>) then
        <address> = <start_address>
        for i = 0 to 14
            if <register_list>[i] == 1
                    Ri_usr = Memory[<address>,4]
                    <address> = <address> + 4
            assert <end_address> == <address> - 4
```

**Exceptions**    Data Abort

**Qualifiers**    Condition Code

**Notes**

**Addressing mode:** The P and U bits distinguish between the different types of addressing mode. See *Addressing Mode 4* starting on page 3-116.

**Banked registers:** LDM must not be followed by an instruction which accesses banked registers (a following NOP is a good way to ensure this).

**Writeback:** Setting bit 21 (the W bit) has UNPREDICTABLE results.

**User and System mode:** LDM is UNPREDICTABLE in user mode or system mode.

**Register Rn:** Specifies the base register used by `<addressing_mode>`.

**Use of R15:** If register 15 if specified as the base register Rn, the result is UNPREDICTABLE.

**Base register mode:** The base register is read from the current processor mode registers, not the user mode registers.

**Data Abort:** If a data abort is signalled, the value left in Rn is the original base register value.

**Non-word-aligned addresses:** LDM instructions ignore the least-significant two bits of `<address>` (words are not rotated as for load word).

**Alignment:** If an implementation includes a System Control Coprocessor (see *Chapter 7*), and alignment checking is enabled, an address with bits[1:0] != 0b00 will cause an alignment exception.

**ARM Architecture Reference Manual**

ARM DUI 0100B

```
LDM{<cond>}<addressing_mode>  Rn{!}, <registers_and_pc>^
```

## LDM (3) ARM

Load and store multiple

Addressing mode 4

**Description**

This form of the LDM (Load Multiple) instruction is useful for returning from an exception, to restore saved registers, load the PC with the return address, update the stack pointer, and restore the CPSR from the SPSR.

In this case, LDM loads a non-empty subset (or possibly all) of the general-purpose registers and the PC from sequential memory locations. The registers are loaded in sequence, the lowest-numbered register first, from the lowest memory address; the highest-numbered register last, from the highest memory address. The SPSR of the current mode is copied to the CPSR.

The instruction is only executed if the condition specified in the instruction matches the condition code status. The conditions are defined in *3.3 The Condition Field* on page 3-4.

| 31    28 | 27 26 25 | 24 | 23 | 22 | 21 | 20 | 19    16 | 15 | 0 |
|----------|----------|----|----|----|----|----|----------|----|---|
| cond | 1 0 0 | P | U | 1 | W | 1 | Rn | 1 | register list |

**Operation**

```
if ConditionPassed(<cond>) then
        <address> = <start_address>
        for i = 0 to 15
              if <register_list>[i] == 1
                        Ri = Memory[<address>,4]
                        <address> = <address> + 4
        assert <end_address> == <address> - 4
        CPSR = SPSR
```

**Exceptions**    Data Abort

**Qualifiers**    Condition Code
! sets the W bit, causing base register update

**Notes**

**Addressing mode:** The P, U and W bits distinguish between the different types of addressing mode. See *Addressing Mode 4* starting on page 3-116.

**Register Rn:** Specifies the base register used by <addressing_mode>.

**Use of R15:** Using R15 as the base register Rn gives an UNPREDICTABLE result.

**User and System mode:** This instruction is UNPREDICTABLE in user or system mode.

**Operand restrictions:** If the base register Rn is specified in <register_list>, and writeback is specified, the final value of Rn is UNPREDICTABLE.

**Data Abort:** If a data abort is signalled and <addressing_mode> specifies writeback, the value left in Rn is IMPLEMENTATION DEFINED, but is either the original base register value or the updated base register value (even if Rn is specified in <register_list>). If register 15 is specified in <register_list>, it must not be overwritten if a data abort occurs.

**Non-word-aligned addresses:** Load multiple instructions ignore the least-significant two bits of <address> (the words are not rotated as for load word).

**Alignment:** If an implementation includes a System Control Coprocessor (see *Chapter 7*), and alignment checking is enabled, an address with bits[1:0] != 0b00 will cause an alignment exception.

# LDR

**Load and store**

**Addressing mode 2**

`LDR{<cond>}  Rd, <addressing_mode>`

**Description**

Combined with a suitable addressing mode, the LDR (Load register) instruction allows 32-bit memory data to be loaded into a general-purpose register where its value can be manipulated. If the destination register is the PC, this instruction loads a 32-bit address from memory and branches to that address (precede the LDR instruction with `MOV LR, PC` to synthesize a branch and link).

Using the PC as the base register allows PC-relative addressing, to facilitate position-independent code.

LDR loads a word from the memory address calculated by `<addressing_mode>` and writes it to register Rd. If the address is not word-aligned, the loaded data is rotated so that the addressed byte occupies the least-significant byte of the register. If the PC is specified as register Rd, the instruction loads a branch to the address (data) into the PC.

The instruction is only executed if the condition specified in the instruction matches the condition code status. The conditions are defined in *3.3 The Condition Field* on page 3-4.

| 31      28 | 27 26 | 25 | 24 | 23 | 22 | 21 | 20 | 19     16 | 15     12 | 11                          0 |
|------------|-------|----|----|----|----|----|----|-----------|-----------|-------------------------------|
| cond       | 0  1  | I  | P  | U  | 0  | W  | 1  | Rn        | Rd        | addressing mode specific      |

**Operation**

```
if ConditionPassed(<cond>) then
    if <address>[1:0] == 0b00
        Rd = Memory[<address>,4]
    else if <address>[1:0] == 0b01
        Rd = Memory[<address>,4] Rotate_Right 8
    else if <address>[1:0] == 0b10
        Rd = Memory[<address>,4] Rotate_Right 16
    else /* <address>[1:0] == 0b11 */
        Rd = Memory[<address>,4] Rotate_Right 24
```

**Exceptions**

Data Abort

**Qualifiers**

Condition Code

**Notes**

**Addressing modes:** The I, P, U and W bits specify the type of `<addressing_mode>` (see *Addressing Mode 2* starting on page 3-98).

**Register Rn:** Specifies the base register used by `<addressing_mode>`.

**Data Abort:** If a data abort is signalled and `<addressing_mode>` uses pre-indexed or post-indexed addressing, the value left in Rn is IMPLEMENTATION DEFINED, but is either the original base register value or the updated base register value (even if the same register is specified for Rd and Rn).

**Operand restrictions:** If `<addressing_mode>` uses pre-indexed or post-indexed addressing, and the same register is specified for Rd and Rn, the results are UNPREDICTABLE.

**Alignment:** If an implementation includes a System Control Coprocessor (See *Chapter 7*), and alignment checking is enabled, an address with bits[1:0] != 0b00 will cause an alignment exception.

**ARM Architecture Reference Manual**

ARM DUI 0100B

**Description**  Combined with a suitable addressing mode, the LDRB (Load Register Byte) instruction allows 8-bit memory data to be loaded into a general-purpose register where it can be manipulated. Using the PC as the base register allows PC-relative addressing, to facilitate position-independent code.

LDRB loads a byte from the memory address calculated by `<addressing_mode>`, zero-extends the byte to a 32-bit word, and writes the word to register Rd.

The instruction is only executed if the condition specified in the instruction matches the condition code status. The conditions are defined in *3.3 The Condition Field* on page 3-4.

| 31 28 | 27 26 25 | 24 | 23 | 22 | 21 | 20 19 | 16 15 | 12 11 | 0 |
|---|---|---|---|---|---|---|---|---|---|
| cond | 0 1 I | P | U | 1 | W | 1 | Rn | Rd | addressing mode specific |

**Operation**
```
if ConditionPassed(<cond>) then
        Rd = Memory[<address>,1]
```

**Exceptions**  Data Abort

**Qualifiers**  Condition Code

**Notes**  **Addressing modes:** The I, P, U and W bits specify the type of `<addressing_mode>` (see *Addressing Mode 2* starting on page 3-98).

**Register Rn:** Specifies the base register used by `<addressing_mode>`.

**Use of R15:** If register 15 is specified for Rd the result is UNPREDICTABLE.

**Operand restrictions:** If `<addressing_mode>` uses pre-indexed or post-indexed addressing, and the same register is specified for Rd and Rn, the results are UNPREDICTABLE.

**Non-word-aligned addresses:** Store Word instructions ignore the least-significant two bits of `<address>` (the words are not rotated as for Load Word).

**Data Abort:** If a data abort is signalled and `<addressing_mode>` uses pre-indexed or post-indexed addressing, the value left in Rn is IMPLEMENTATION DEFINED, but is either the original base register value or the updated base register value (even if the same register is specified for Rd and Rn).

LDR{<cond>}BT   Rd, <post_indexed_addressing_mode>

*Load and store*

*Addressing mode 2*

**Description**   The LDRBT (Load Register Byte with Translation) instruction can be used by a (privileged) exception handler that is emulating a memory access instruction that would normally execute in User Mode. The access is restricted as if it has User Mode privilege.

LDRBT loads a byte from the memory address calculated by <post_indexed_addressing_mode>, zero-extends the byte to a 32-bit word, and writes the word to register Rd. If the instruction is executed when the processor is in a privileged mode, the memory system is signalled to treat the access as if the processor was in user mode.

The instruction is only executed if the condition specified in the instruction matches the condition code status. The conditions are defined in *3.3 The Condition Field* on page 3-4.

| 31   28 | 27 | 26 | 25 | 24 | 23 | 22 | 21 | 20 | 19   16 | 15   12 | 11   0 |
|---------|----|----|----|----|----|----|----|----|---------|---------|--------|
| cond | 0 | 1 | I | 0 | U | 1 | 1 | 1 | Rn | Rd | addressing mode specific |

**Operation**
```
if ConditionPassed(<cond>) then
        Rd = Memory[<address>,1]
```

**Exceptions**   Data Abort

**Qualifiers**   Condition Code

**Notes**   **Addressing modes:** The I, P, and U bits specify the type of <addressing_mode> (see *Addressing Mode 2* starting on page 3-98).

**Register Rn:** Specifies the base register used by <post_indexed_addressing_mode>.

**User mode:** If this instruction is executed in user mode, an ordinary user mode access is performed.

**Use of R15:** If register 15 is specified for Rd, the result is UNPREDICTABLE.

**Operand restrictions:** If the same register is specified for Rd and Rn, the results are UNPREDICTABLE.

**Data Abort:** If a data abort is signalled, the value left in Rn is IMPLEMENTATION DEFINED, but is either the original base register value or the updated base register value (even if the same register is specified for Rd and Rn).

**ARM Architecture Reference Manual**

ARM DUI 0100B

**Description**   Used with a suitable addressing mode, the LDRH (Load Register Halfword) instruction allows 16-bit memory data to be loaded into a general-purpose register where its value can be manipulated.

Using the PC as the base register allows PC-relative addressing to facilitate position-independent code.

LDRH loads a halfword from the memory address calculated by <addressing_mode>, zero-extends the halfword to a 32-bit word, and writes the word to register Rd. If the address is not halfword-aligned, the result is UNPREDICTABLE.

The instruction is only executed if the condition specified in the instruction matches the condition code status. The conditions are defined in *3.3 The Condition Field* on page 3-4.

| 31    28 | 27 26 25 | 24 | 23 | 22 | 21 | 20 | 19        16 | 15      12 | 11        8 | 7 | 6 | 5 | 4 | 3        0 |
|----------|----------|----|----|----|----|----|--------------|------------|-------------|---|---|---|---|------------|
| cond     | 0  0  0  | P  | U  | I  | W  | 1  | Rn           | Rd         | addr_mode   | 1 | 0 | 1 | 1 | addr_mode  |

**Operation**
```
if ConditionPassed(<cond>) then
    if <address>[0] == 0
        <data> = Memory[<address>,2]
    else /* <address>[0] == 1 */
        <data> = UNPREDICTABLE
    Rd = <data>
```

**Exceptions**   Data Abort

**Qualifiers**   Condition Code

**Notes**   **Addressing modes:** The I, P, U and W bits specify the type of <addressing_mode> (see *Addressing Mode 3* starting on page 3-109).

**The addr_mode bits:** These bits are addressing-mode specific.

**Register Rn:** Specifies the base register used by <addressing_mode>.

**Use of R15:** If register 15 is specified for Rd, the result is UNPREDICTABLE.

**Operand restrictions:** If <addressing_mode> uses pre-indexed or post-indexed addressing, and the same register is specified for Rd and Rn, the results are UNPREDICTABLE.

**Data Abort:** If a data abort is signalled and <addressing_mode> uses pre-indexed or post-indexed addressing, the value left in Rn is IMPLEMENTATION DEFINED, but is either the original base register value or the updated base register value (even if the same register is specified for Rd and Rn).

**Non-half-word aligned addresses:** If the load address is not halfword-aligned, the loaded value is UNPREDICTABLE.

**Alignment:** If an implementation includes a System Control Coprocessor (see *Chapter 7*), and alignment checking is enabled, an address with bit[0] != 0 will cause an alignment exception.

**ARM Architecture Reference Manual**
ARM DUI 0100B

Load and store

Addressing mode 3

Architecture v4 only

`LDR{<cond>}SB  Rd, <addressing_mode>`

**Description**    Used with a suitable addressing mode, the LDRSB (Load Register Signed Byte) instruction allows 8-bit signed memory data to be loaded into a general-purpose register where it can be manipulated.

Using the PC as the base register allows PC-relative addressing, to facilitate position-independent code.

LDRSB loads a byte from the memory address calculated by `<addressing_mode>`, sign extends the byte to a 32-bit word, and writes the word to register Rd.

The instruction is only executed if the condition specified in the instruction matches the condition code status. The conditions are defined in *3.3 The Condition Field* on page 3-4.

| 31 28 | 27 26 25 | 24 | 23 | 22 | 21 | 20 | 19 16 | 15 12 | 11 8 | 7 6 5 4 | 3 0 |
|---|---|---|---|---|---|---|---|---|---|---|---|
| cond | 0 0 0 | P | U | I | W | 1 | Rn | Rd | addr_mode | 1 1 0 1 | addr_mode |

**Operation**
```
if ConditionPassed(<cond>) then
        <data> = Memory[<address>,1]
        Rd = SignExtend(<data>)
```

**Exceptions**    Data Abort

**Qualifiers**    Condition Code

**Notes**    **Addressing modes:** The I, P, U and W bits specify the type of `<addressing_mode>` (see *Addressing Mode 3* starting on page 3-109).

**The addr_mode bits:** These bits are addressing mode specific.

**Register Rn:** Specifies the base register used by `<addressing_mode>`.

**Use of R15:** If register 15 is specified for Rd, the result is UNPREDICTABLE.

**Operand restrictions:** If `<addressing_mode>` uses pre-indexed or post-indexed addressing, and the same register is specified for Rd and Rn, the results are UNPREDICTABLE.

**Data Abort:** If a data abort is signalled and `<addressing_mode>` uses pre-indexed or post-indexed addressing, the value left in Rn is IMPLEMENTATION DEFINED, but is either the original base register value or the updated base register value (even if the same register is specified for Rd and Rn).

**ARM Architecture Reference Manual**

ARM DUI 0100B

```
LDR{<cond>}SH   Rd, <addressing_mode>
```

**Description**   Used with a suitable addressing mode, the LDRSH (Load Register Signed Halfword) instruction allows 16-bit signed memory data to be loaded into a general-purpose register where its value can be manipulated.

Using the PC as the base register allows PC-relative addressing, to facilitate position-independent code.

LDRSH loads a halfword from the memory address calculated by <addressing_mode>, sign-extends the halfword to a 32-bit word, and writes the word to register Rd. If the address is not halfword-aligned, the result is UNPREDICTABLE.

The instruction is only executed if the condition specified in the instruction matches the condition code status. The conditions are defined in *3.3 The Condition Field* on page 3-4.

| 31      28 | 27 26 25 | 24 | 23 | 22 | 21 | 20 | 19      16 | 15      12 | 11      8 | 7 | 6 | 5 | 4 | 3      0 |
|---|---|---|---|---|---|---|---|---|---|---|---|---|---|---|
| cond | 0 0 0 | P | U | I | W | 1 | Rn | Rd | addr_mode | 1 | 1 | 1 | 1 | addr_mode |

**Operation**
```
if ConditionPassed(<cond>) then
    if <address>[0] == 0
        <data> = Memory[<address>,2]
    else /* <address>[0] == 1 */
        <data> = UNPREDICTABLE
    Rd = SignExtend(<data>)
```

**Exceptions**   Data Abort

**Qualifiers**   Condition Code

**Notes**   **Addressing modes:** The I, P, U and W bits specify the type of <addressing_mode> (see *Addressing Mode 3* starting on page 3-109).

**The addr_mode bits:** These bits are addressing mode specific.

**Register Rn:** Specifies the base register used by <addressing_mode>.

**Use of R15:** If register 15 is specified for Rd, the result is UNPREDICTABLE.

**Operand restrictions:** If <addressing_mode> uses pre-indexed or post-indexed addressing, and the same register is specified for Rd and Rn, the results are UNPREDICTABLE.

**Data Abort:** If a data abort is signalled and <addressing_mode> uses pre-indexed or post-indexed addressing, the value left in Rn is IMPLEMENTATION DEFINED, but is either the original base register value or the updated base register value (even if the same register is specified for Rd and Rn).

**Non-half-word aligned addresses:** If the load address is not halfword-aligned, the loaded value is UNPREDICTABLE.

**Alignment:** If an implementation includes a System Control Coprocessor (see *Chapter 7*), and alignment checking is enabled, an address with bit[0] != 0 causes an alignment exception.

**ARM Architecture Reference Manual**
ARM DUI 0100B

`LDR{<cond>}T  Rd, <post_indexed_addressing_mode>`

**Description**

The LDRT (Load Register with Translation) instruction can be used by a (privileged) exception handler that is emulating a memory access instruction that would normally execute in User Mode. The access is restricted as if it has User Mode privilege.

LDRT loads a word from the memory address and writes it to register Rd. If the instruction is executed when the processor is in a privileged mode, the memory system is signalled to treat the access as if the processor was in user mode.

The instruction is only executed if the condition specified in the instruction matches the condition code status. The conditions are defined in *3.3 The Condition Field* on page 3-4.

| 31 28 | 27 26 25 24 23 22 21 20 | 19 16 | 15 12 | 11 0 |
|---|---|---|---|---|
| cond | 0 1 I 0 U 0 1 1 | Rn | Rd | addressing mode specific |

**Operation**

```
if ConditionPassed(<cond>) then
    if <address>[1:0] == 0b00
        Rd = Memory[<address>,4]
    else if <address>[1:0] == 0b01
        Rd = Memory[<address>,4] Rotate_Right 8
    else if <address>[1:0] == 0b10
        Rd = Memory[<address>,4] Rotate_Right 16
    else /* <address>[1:0] == 0b11 */
        Rd = Memory[<address>,4] Rotate_Right 24
```

**Exceptions**    Data Abort

**Qualifiers**    Condition Code

**Notes**

**Addressing modes:** The I, P, and U bits specify the type of `<addressing_mode>` (see *Addressing Mode 2* starting on page 3-98).

**Register Rn:** Specifies the base register used by `<post_indexed_addressing_mode>`.

**User mode:** If this instruction is executed in user mode, an ordinary user mode access is performed.

**Operand restrictions:** If the same register is specified for Rd and Rn the results are UNPREDICTABLE.

**Data Abort:** If data abort is signalled, the value left in Rn is IMPLEMENTATION DEFINED, but is either the original base register value or the updated base register value (even if the same register is specified for Rd and Rn).

**Alignment:** If an implementation includes a System Control Coprocessor (See *Chapter 7*), and alignment checking is enabled, an address with bits[1:0] != 0b00 causes an alignment exception.

**ARM Architecture Reference Manual**

ARM DUI 0100B

**Description**   The MCR (Move to Coprocessor from ARM Register) instruction is used to initiate coprocessor instructions that operate on values in ARM registers, for example a fixed-point to floating-point conversion instruction for a floating-point accelerator coprocessor.

MCR passes the value of register Rd to the coprocessor specified by <cp_num>. If no coprocessors indicate that they can execute the instruction, an UNDEFINED instruction exception is generated.

*Not in architecture v1*

The instruction is only executed if the condition specified in the instruction matches the condition code status. The conditions are defined in *3.3 The Condition Field* on page 3-4.

| 31      28 | 27 26 25 24 | 23     21 | 20 | 19      16 | 15      12 | 11      8 | 7      5 | 4 | 3      0 |
|---|---|---|---|---|---|---|---|---|---|
| cond | 1 1 1 0 | opcode_1 | 0 | CRn | Rd | cp_num | opcode_2 | 1 | CRm |

**Operation**   
```
if ConditionPassed(<cond>) then
        Coprocessor[<cp_num>] = Rd
```

**Exceptions**   Undefined Instruction

**Qualifiers**   Condition Code

**Notes**   **Coprocessor fields:**  Only instruction bits[31:24], bit[20], bits[15:8] and bit[4] are ARM architecture-defined; the remaining fields are only recommendations, for compatibility with ARM Development Systems.

**ARM Architecture Reference Manual**
ARM DUI 0100B

MLA{<cond>}{<S>}  Rd, Rm, Rs, Rn

*Multiply*

*Not in architecture v1*

**Description**   The MLA (Multiply Accumulate) instruction multiplies signed or unsigned operands to produce a 32-bit result, which is then added to a third operand, and written to the destination register.

MLA multiplies the value of register Rm with the value of register Rs, adds the value of register Rn, and stores the result in the destination register Rd. The condition code flags are optionally updated (based on the result).

The instruction is only executed if the condition specified in the instruction matches the condition code status. The conditions are defined in *3.3 The Condition Field* on page 3-4.

| 31    28 | 27 26 25 24 23 22 21 20 | 19    16 | 15    12 | 11    8 | 7 6 5 4 | 3    0 |
|----------|-------------------------|----------|----------|---------|---------|--------|
| cond     | 0 0 0 0 0 0 1 S         | Rd       | Rn       | Rs      | 1 0 0 1 | Rm     |

**Operation**
```
if ConditionPassed(<cond>) then
    Rd = (Rm * Rs + Rn)[31:0]
    if S == 1 then
        N Flag = Rd[31]
        Z Flag = if Rd == 0 then 1 else 0
        C Flag = UNPREDICTABLE
        V Flag = unaffected
```

**Exceptions**   None

**Qualifiers**   Condition Code
S updates condition code flags N and Z

**Notes**   **Use of R15:** Specifying R15 for register Rd, Rm, Rs or Rn has UNPREDICTABLE results.

**Operand restriction:** Specifying the same register for Rd and Rm has UNPREDICTABLE results.

**Early termination:** If the multiplier implementation supports early termination, it must be implemented on the value of the Rs operand. The type of early termination used (signed or unsigned) is IMPLEMENTATION DEFINED.

**Signed and unsigned:** Because the MLA instruction produces only the lower 32 bits of the 64-bit product, MLA gives the same answer for multiplication of both signed and unsigned numbers.

**ARM Architecture Reference Manual**

ARM DUI 0100B

**Description**  The MOV (Move) instruction is used to:

- move a value from one register to another
- put a constant value into a register
- perform a shift without any other arithmetic or logical operation

When the PC is the destination of the instruction, a branch occurs, and `MOV PC, LR` can be used to return from a subroutine call (see the B and BL instructions) and to return from some types of exception (See *2.5 Exceptions* on page 2-6).

MOV moves the value of `<shifter_operand>` to the destination register Rd, and optionally updates the condition code flags (based on the result).

The instruction is only executed if the condition specified in the instruction matches the condition code status. The conditions are defined in *3.3 The Condition Field* on page 3-4.

| 31    28 | 27 26 | 25 | 24 23 22 21 | 20 | 19        16 | 15      12 | 11                    0 |
|----------|-------|----|-------------|----|--------------|------------|-------------------------|
| cond | 0 0 | I | 1 1 0 1 | S | SBZ | Rd | shifter_operand |

**Operation**
```
if ConditionPassed(<cond>) then
        Rd = <shifter_operand>
        if S == 1 and Rd == R15 then
           CPSR = SPSR
        else if S == 1 then
           N Flag = Rd[31]
           Z Flag = if Rd == 0 then 1 else 0
           C Flag = <shifter_carry_out>
           V Flag = unaffected
```

**Exceptions**  None

**Qualifiers**  Condition Code
S updates condition code flags N,Z and C

**Notes**  **Shifter operand:** The shifter operands for this instruction are given in *Addressing Mode 1* starting on page 3-84.

**Writing to R15:** When Rd is R15 and the S flag in the instruction is not set, the result of the operation is placed in the PC. When Rd is R15 and the S flag is set, the result of the operation is placed in the PC and the SPSR corresponding to the current mode is moved to the CPSR. This allows state changes which atomically restore both PC and CPSR. This form of the instruction is UNPREDICTABLE in User mode and System mode.

**The I bit:** Bit 25 is used to distinguish between the immediate and register forms of `<shifter_operand>`.

**ARM Architecture Reference Manual**
ARM DUI 0100B

*Load and store multiple*

*Addressing mode 4*

**Description**

The MRC (Move to ARM Register from Coprocessor) instruction is used to initiate coprocessor instructions that return values to ARM registers, for example a floating-point to fixed-point conversion instruction for a floating-point accelerator coprocessor.

Specifying R15 as the destination register is useful for operations like a floating-point compare instruction.

MRC has two uses:

1    If Rd specifies register 15, the condition code flags bits are updated from the top four bits of the value from the coprocessor specified by <cp_num> (to allow conditional branching on the status of a coprocessor) and the other 28 bits are IGNORED.

2    Otherwise the instruction writes into register Rd a value from the coprocessor specified by <cp#>.

If no coprocessors indicate that they can execute the instruction an UNDEFINED instruction exception is generated.

The instruction is only executed if the condition specified in the instruction matches the condition code status. The conditions are defined in *3.3 The Condition Field* on page 3-4.

| 31      28 | 27 26 25 24 | 23      21 | 20 | 19      16 | 15      12 | 11      8 | 7      5 | 4 | 3      0 |
|---|---|---|---|---|---|---|---|---|---|
| cond | 1  1  1  0 | opcode_1 | 1 | CRn | Rd | cp_num | opcode_2 | 1 | CRm |

**Operation**

```
if ConditionPassed(<cond>) then
    if Rd == 15 then
        N flag = (value from Coprocessor[<cp_num>])[31]
        Z flag = (value from Coprocessor[<cp_num>])[30]
        C flag = (value from Coprocessor[<cp_num>])[29]
        V flag = (value from Coprocessor[<cp_num>])[28]
    else /* Rd != 15 */
        Rd = value from Coprocessor[<cp_num>]
```

**Exceptions**    Undefined Instruction

**Qualifiers**    Condition Code

**Notes**    **Coprocessor fields:** Only instruction bits[31:24], bit[20], bits[15:8] and bit[4] are ARM architecture-defined; the remaining fields are only recommendations, that if followed, will be compatible with ARM Development Systems.

**ARM Architecture Reference Manual**

ARM DUI 0100B

```
MRS{<cond>} Rd, CPSR
MRS{<cond>} Rd, SPSR
```

*Arch 3 & 4 only*

**Description**    The MRS instruction moves the value of the CPSR or the SPSR of the current mode into a general-purpose register. In the general-purpose register, the value can be examined or manipulated with normal data-processing instructions.

The MRS moves the value of the CPSR, or the value of the SPSR corresponding to the current mode, to a general-purpose register.

The instruction is only executed if the condition specified in the instruction matches the condition code status. The conditions are defined in *3.3 The Condition Field* on page 3-4.

| 31      28 | 27 26 25 24 23 | 22 | 21 20 | 19        16 | 15      12 | 11                    0 |
|------------|----------------|----|-------|--------------|------------|-------------------------|
| cond | 0  0  0  1  0 | R | 0  0 | SBO | Rd | SBZ |

**Operation**    
```
if ConditionPassed(<cond>) then
        if R == 1 then
            Rd = SPSR
        else
            Rd = CPSR
```

**Exceptions**    None

**Qualifiers**    Condition Code

**Notes**    **Opcode [11:0]:** Execution of MRS instructions with any non-zero bits in opcode[11:0] is UNPREDICTABLE.

**Opcode [19:16]:** Execution of MRS instructions with any non-one bits in opcode[19:16] is UNPREDICTABLE.

**User mode SPSR:** Accessing the SPSR when in user mode or system mode is UNPREDICTABLE.

*Contradicts page
3-24 for V.4*

*R15 as destination?*

```
MSR{<cond>}   CPSR_f, #32bit immediate
MSR{<cond>}   CPSR_<fields>,Rm
MSR{<cond>}   SPSR_f, #32bit immediate
MSR{<cond>}   SPSR_<fields>, Rm
```

*Status register access*

*Architecture v3
and v4 only*

**Description**  The MSR (Move to Status register from ARM Register) instruction transfers the value of a general-purpose register to the CPSR or the SPSR of the current mode. This is used to update the value of the condition code flags, interrupt enables, or the processor mode.

MSR moves the value of Rm or the value of the 32-bit immediate (encoded as an 8-bit value with rotate) to the CPSR or the SPSR corresponding to the current mode.

The instruction is only executed if the condition specified in the instruction matches the condition code status. The conditions are defined in *3.3 The Condition Field* on page 3-4.

**Immediate operand**

| 31    28 | 27 26 25 24 23 | 22 | 21 20 | 19    16 | 15    12 | 11    8 | 7    0 |
|----------|----------------|----|-------|----------|----------|---------|--------|
| cond | 0 0 1 1 0 | R | 1 0 | field_mask | SBO | rotate_imm | 8_bit_immediate |

**Register operand**

| 31    28 | 27 26 25 24 23 | 22 | 21 20 | 19    16 | 15    12 | 11    5 | 4 | 3    0 |
|----------|----------------|----|-------|----------|----------|---------|---|--------|
| cond | 0 0 0 1 0 | R | 1 0 | field_mask | SBO | SBZ | 0 | Rm |

**Operation**

```
if ConditionPassed(<cond>) then
    if opcode[25] == 1
        <operand> = <8_bit_immediate> Rotate_Right (<rotate_imm> * 2)
    else /* opcode[25] == 0 */
        <operand>
    if R == 0 then
        if <field_mask>[0] == 1 and InAPrivilegedMode() then
            CPSR[7:0] = <operand>[7:0]
        if <field_mask>[1] == 1 and InAPrivilegedMode() then
            CPSR[15:8] = <operand>[15:8]
        if <field_mask>[2] == 1 and InAPrivilegedMode() then
            CPSR[23:16] = <operand>[23:16]
        if <field_mask>[3] == 1 then
            CPSR[31:24] = <operand>[31:24]
    else /* R == 1 */
        if <field_mask>[0] == 1 and CurrentModeHasSPSR() then
            SPSR[7:0] = <operand>[7:0]
        if <field_mask>[1] == 1 and CurrentModeHasSPSR() then
            SPSR[15:8] = <operand>[15:8]
        if <field_mask>[2] == 1 and CurrentModeHasSPSR() then
            SPSR[23:16] = <operand>[23:16]
        if <field_mask>[3] == 1 and CurrentModeHasSPSR() then
            SPSR[31:24] = <operand>[31:24]
```

**ARM Architecture Reference Manual**

ARM DUI 0100B

**Exceptions**      None

**Qualifiers**      Condition Code

<fields> is one of

    _c      sets the control field mask bit (bit 0)

    _x      sets the extension field mask bit (bit 1)

    _s      sets the status field mask bit (bit 2)

    _f      sets the flags field mask bit (bit 3)

*Architecture v3
and v4 only*

**Notes**

**Immediate Operand:** The immediate form of this instruction can only be used to set the flag bits (PSR bits 31:24). Using the immediate form on any other fields has UNPREDICTABLE results.

**PSR Update:** The value of a PSR must be updated by moving the PSR to a general-purpose register (using the MRS instruction), modifying the relevant bits of the general-purpose register, and restoring the updated general-purpose register value back into the PSR (using the MSR instruction).

**User Mode CPSR:** Any writes to CPSR[23:0] in user mode are IGNORED (so that user mode programs cannot change to a privileged mode).

**User mode SPSR:** Accessing the SPSR when in user mode is UNPREDICTABLE.

**System mode SPSR:** Accessing the SPSR when in system mode is UNPREDICTABLE.

**Deprecated field specification:** The CPSR, CPSR_flg, CPSR_ctl, CPSR_all, SPSR, SPSR_flg, SPSR_ctl and SPSR_all forms of PSR field specification have been superseded by the csxf format shown above.

CPSR, SPSR, CPSR_all and SPSR_all produce a field mask of 0b1001.
CPSR_flg and SPSR_flg produce a field mask of 0b1000.
CPSR_ctl and SPSR_ctl produce a field mask of 0b0001.

# MUL

*Multiply*

*Not in architecture v1*

MUL{<cond>}{<S>}  Rd, Rm, Rs

**Description**

The MUL (Multiply) instruction is used to multiply signed or unsigned variables to produce a 32-bit result.

MUL multiplies the value of register Rm with the value of register Rs, and stores the result in the destination register Rd. The condition code flags are optionally updated (based on the result).

The instruction is only executed if the condition specified in the instruction matches the condition code status. The conditions are defined in *3.3 The Condition Field* on page 3-4.

| 31 | 28 | 27 26 25 24 23 22 21 | 20 | 19 | 16 | 15 | 12 | 11 | 8 | 7 6 5 4 | 3 | 0 |
|---|---|---|---|---|---|---|---|---|---|---|---|---|
| cond | | 0 0 0 0 0 0 0 | S | Rn | | SBZ | | Rs | | 1 0 0 1 | Rm | |

**Operation**

```
if ConditionPassed(<cond>) then
    Rd = (Rm * Rs)[31:0]
    if S == 1 then
        N Flag = Rd[31]
        Z Flag = if Rd == 0 then 1 else 0
        C Flag = UNPREDICTABLE
        V Flag = unaffected
```

*oops!*

**Exceptions**    None

**Qualifiers**    Condition Code
S update condition code flags N,Z

**Notes**

**Use of R15:** Specifying R15 for register Rd, Rm or Rs has UNPREDICTABLE results.

**Operand restriction:** Specifying the same register for Rd and Rm has UNPREDICTABLE results.

**Early termination:** If the multiplier implementation supports early termination, it must be implemented on the value of the Rs operand. The type of early termination used (signed or unsigned) is IMPLEMENTATION DEFINED.

**Signed and unsigned:** Because the MUL instruction produces only the lower 32 bits of the 64-bit product, MUL gives the same answer for multiplication of both signed and unsigned numbers.

**ARM Architecture Reference Manual**

ARM DUI 0100B

**Description**    The MVN (Move negative) instruction is used to:

- write a negative value into a register
- form a bit mask
- take the one's complement of a value

MVN moves the logical one's compliment of the value of `<shifter_operand>` to the destination register Rd, and optionally updates the condition code flags (based on the result).

The instruction is only executed if the condition specified in the instruction matches the condition code status. The conditions are defined in *3.3 The Condition Field* on page 3-4.

| 31    28 | 27 26 | 25 | 24 23 22 21 | 20 | 19    16 | 15    12 | 11    0 |
|----------|-------|----|-------------|----|----------|----------|---------|
| cond | 0  0 | I | 1  1  1  1 | S | SBZ | Rd | shifter_operand |

**Operation**
```
if ConditionPassed(<cond>) then
    Rd = NOT <shifter_operand>
    if S == 1 and Rd == R15 then
        CPSR = SPSR
    else if S == 1 then
        N Flag = Rd[31]
        Z Flag = if Rd == 0 then 1 else 0
        C Flag = <shifter_carry_out>
        V Flag = unaffected
```

**Exceptions**    None

**Qualifiers**    Condition Code
S                    Update condition code flags N,Z,C

**Notes**    **Shifter operand:** The shifter operands for this instruction are given in *Addressing Mode 1* starting on page 3-84.

**Writing to R15:** When Rd is R15 and the S flag in the instruction is not set, the result of the operation is placed in the PC. When Rd is R15 and the S flag is set, the result of the operation is placed in the PC and the SPSR corresponding to the current mode is moved to the CPSR. This allows state changes which atomically restore both PC and CPSR. This form of the instruction is UNPREDICTABLE in User mode and System mode.

**The I bit:** Bit 25 is used to distinguish between the immediate and register forms of `<shifter_operand>`.

`ORR{<cond>}{S}  Rd, Rn, <shifter_operand>`

*Data processing*

*Addressing mode 1*

**Description**   The ORR (Logical OR) instruction can be used to set selected bits in a register; for each bit OR with 1 will set the bit, OR with 0 will leave it unchanged.

ORR performs a bitwise (inclusive) OR of the value of register Rn with the value of `<shifter_operand>`, and stores the result in the destination register Rd. The condition code flags are optionally updated (based on the result).

The instruction is only executed if the condition specified in the instruction matches the condition code status. The conditions are defined in *3.3 The Condition Field* on page 3-4.

| 31    28 | 27 26 | 25 | 24 23 22 21 | 20 | 19    16 | 15    12 | 11    0 |
|----------|-------|----|-------------|----|----------|----------|---------|
| cond | 0  0 | I | 1  1  0  0 | S | Rn | Rd | shifter_operand |

**Operation**
```
if ConditionPassed(<cond>) then
    Rd = Rn OR <shifter_operand>
    if S == 1 and Rd == R15 then
        CPSR = SPSR
    else if S == 1 then
        N Flag = Rd[31]
        Z Flag = if Rd == 0 then 1 else 0
        C Flag = <shifter_carry_out>
        V Flag = unaffected
```

**Exceptions**   None

**Qualifiers**   Condition Code
S updates condition code flags N, Z and C

**Notes**   **Shifter operand:** The shifter operands for this instruction are given in *Addressing Mode 1* starting on page 3-84.

**Writing to R15:** When Rd is R15 and the S flag in the instruction is not set, the result of the operation is placed in the PC. When Rd is R15 and the S flag is set, the result of the operation is placed in the PC and the SPSR corresponding to the current mode is moved to the CPSR. This allows state changes which atomically restore both PC and CPSR. This form of the instruction is UNPREDICTABLE in User mode and System mode.

**The I bit:** Bit 25 is used to distinguish between the immediate and register forms of `<shifter_operand>`.

**ARM Architecture Reference Manual**
ARM DUI 0100B

**Description**   The RSB (Reverse Subtract) instruction subtracts the value of register Rn from the value of <shifter_operand>, and stores the result in the destination register Rd. The condition code flags are optionally updated (based on the result).

The following instruction stores the negative (two's complement) of Rx in Rd.

```
RSB Rd, Rx, #0
```

Constant multiplication (of Rx) by $2^n-1$ (into Rd) can be performed with:

```
RSB Rd, Rx, Rx LSL #n
```

The instruction is only executed if the condition specified in the instruction matches the condition code status. The conditions are defined in *3.3 The Condition Field* on page 3-4.

| 31    28 | 27 26 | 25 | 24 23 22 21 | 20 | 19    16 | 15    12 | 11    0 |
|----------|-------|----|-------------|----|----------|----------|---------|
| cond | 0 0 | I | 0 0 1 1 | S | Rn | Rd | shifter_operand |

**Operation**
```
if ConditionPassed(<cond>) then
    Rd = <shifter_operand> - Rn
    if S == 1 and Rd == R15 then
        CPSR = SPSR
    else if S == 1 then
        N Flag = Rd[31]
        Z Flag = if Rd == 0 then 1 else 0
        C Flag = NOT BorrowFrom(<shifter_operand> - Rn)
        V Flag = OverflowFrom (<shifter_operand> - Rn)
```

**Exceptions**   None

**Qualifiers**   Condition Code
S updates condition code flags N,Z,C,V

**Notes**   **Shifter operand:** The shifter operands for this instruction are given in *Addressing Mode 1* starting on page 3-84.

**Writing to R15:**  When Rd is R15 and the S flag in the instruction is not set, the result of the operation is placed in the PC. When Rd is R15 and the S flag is set, the result of the operation is placed in the PC and the SPSR corresponding to the current mode is moved to the CPSR. This allows state changes which atomically restore both PC and CPSR. This form of the instruction is UNPREDICTABLE in User mode and System mode.

**The I bit:** Bit 25 is used to distinguish between the immediate and register forms of <shifter_operand>.

**ARM Architecture Reference Manual**
ARM DUI 0100B

# RSC

**ARM**

*Data processing*

*Addressing mode 1*

`RSC{<cond>}{S}  Rd, Rn, <shifter_operand>`

**Description**

The RSC (Reverse Subtract with Carry) instruction subtracts the value of register Rn and the value of NOT (Carry Flag) from the value of `<shifter_operand>`, and stores the result in the destination register Rd. The condition code flags are optionally updated (based on the result).

To negate the 64-bit value in R0, R1, use the following sequence (R0 holds the least-significant word) and store the result in R2,R3:

```
RSBS   R2,R0,#0
RSC    R3,R1,#0
```

The instruction is only executed if the condition specified in the instruction matches the condition code status. The conditions are defined in *3.3 The Condition Field* on page 3-4.

| 31  28 | 27 26 | 25 | 24 23 22 21 | 20 | 19  16 | 15  12 | 11  0 |
|---|---|---|---|---|---|---|---|
| cond | 0 0 | I | 0 1 1 1 | S | Rn | Rd | shifter_operand |

**Operation**

```
if ConditionPassed(<cond>) then
    Rd = <shifter_operand> - Rn - NOT(C Flag)
    if S == 1 and Rd == R15 then
        CPSR = SPSR
    else if S == 1 then
        N Flag = Rd[31]
        Z Flag = if Rd == 0 then 1 else 0
        C Flag = NOT BorrowFrom(<shifter_operand> - Rn - NOT(C Flag))
        V Flag = OverflowFrom (<shifter_operand> - Rn - NOT(C Flag))
```

**Exceptions**   None

**Qualifiers**
Condition Code
S updates condition code flags N,Z,C and V

**Notes**

**Shifter operand:** The shifter operands for this instruction are given in *Addressing Mode 1* starting on page 3-84.

**Writing to R15:** When Rd is R15 and the S flag in the instruction is not set, the result of the operation is placed in the PC. When Rd is R15 and the S flag is set, the result of the operation is placed in the PC and the SPSR corresponding to the current mode is moved to the CPSR. This allows state changes which atomically restore both PC and CPSR. This form of the instruction is UNPREDICTABLE in User mode and System mode.

**The I bit:** Bit 25 is used to distinguish between the immediate and register forms of `<shifter_operand>`.

**ARM Architecture Reference Manual**
ARM DUI 0100B

**Description**
The SBC (Subtract with Carry) instruction is used to synthesize multi-word subtraction. If register pairs R0,R1 and R2,R3 hold 64-bit values (R0 and R2 hold the least-significant words), the following instructions leave the 64-bit difference in R4,R5:

```
SUBS   R4,R0,R2
SBC    R5,R1,R3
```

SBC subtracts the value of `<shifter_operand>` and the value of NOT (Carry Flag) from the value of register Rn, and stores the result in the destination register Rd. The condition code flags are optionally updated (based on the result).

The instruction is only executed if the condition specified in the instruction matches the condition code status. The conditions are defined in *3.3 The Condition Field* on page 3-4.

| 31 | | 28 | 27 | 26 | 25 | 24 | 23 | 22 | 21 | 20 | 19 | | 16 | 15 | | 12 | 11 | | 0 |
|----|----|----|----|----|----|----|----|----|----|----|----|----|----|----|----|----|----|----|----|
| cond | | | 0 | 0 | I | 0 | 1 | 1 | 0 | S | Rn | | | Rd | | | shifter_operand | | |

**Operation**
```
if ConditionPassed(<cond>) then
    Rd = Rn - <shifter_operand> - NOT(C Flag)
    if S == 1 and Rd == R15 then
        CPSR = SPSR
    else if S == 1 then
        N Flag = Rd[31]
        Z Flag = if Rd == 0 then 1 else 0
        C Flag = NOT BorrowFrom(Rn - <shifter_operand> - NOT(C Flag))
        V Flag = OverflowFrom (Rn - <shifter_operand> - NOT(C Flag))
```

**Exceptions**     None

**Qualifiers**     Condition Code
S                 Update condition code flags N,Z,C,V

**Notes**          **Shifter operand:** The shifter operands for this instruction are given in *Addressing Mode 1* starting on page 3-84.

**Writing to R15:** When Rd is R15 and the S flag in the instruction is not set, the result of the operation is placed in the PC. When Rd is R15 and the S flag is set, the result of the operation is placed in the PC and the SPSR corresponding to the current mode is moved to the CPSR. This allows state changes which atomically restore both PC and CPSR. This form of the instruction is UNPREDICTABLE in User mode and System mode.

**The I bit:** Bit 25 is used to distinguish between the immediate and register forms of `<shifter_operand>`.

**ARM Architecture Reference Manual**
ARM DUI 0100B

SMLAL{<cond>}{<S>} RdLo, RdHi, Rm, Rs

*Multiply*

**Description**

The SMLAL (Signed Multiply Accumulate Long) instruction multiplies signed variables to produce a 64-bit result, which is added to the 64-bit value in the two destination general-purpose registers. The result is written back to the two destination general-purpose registers.

*Architecture v3 and v4 only*

SMLAL multiplies the signed value of register Rm with the signed value of register Rs to produce a 64-bit result. The lower 32 bits of the result are added to RdLo and stored in RdLo; the upper 32 bits, and the carry from the addition to RdLo, are added to RdHi and stored in RdHi. The condition code flags are optionally updated (based on the 64-bit result).

The instruction is only executed if the condition specified in the instruction matches the condition code status. The conditions are defined in *3.3 The Condition Field* on page 3-4.

| 31      28 | 27 26 25 24 23 22 21 | 20 | 19      16 | 15      12 | 11      8 | 7 6 5 4 | 3      0 |
|------------|---------------------|----|-----------|-----------|----------|---------|----------|
| cond | 0 0 0 0 1 1 1 | S | RdHi | RdLo | Rs | 1 0 0 1 | Rm |

**Operation**

```
if ConditionPassed(<cond>) then
    RdLo = (Rm * Rs)[31:0] + RdLo
    RdHi = (Rm * Rs)[63:32] + RdHi + CarryFrom((Rm * Rs)[31:0] + RdLo)
    if S == 1 then
        N Flag = RdHi[31]
        Z Flag = if (RdHi == 0) and (RdLo == 0) then 1 else 0
        C Flag = UNPREDICTABLE
        V Flag = UNPREDICTABLE
```

**Exceptions**   None

**Qualifiers**   Condition Code
S updates condition code flags N,Z

**Notes**

**Use of R15:** Specifying R15 for register RdHi, RdLo, Rm or Rs has UNPREDICTABLE results.

**Operand restriction:** Specifying the same register for RdHi and Rm has UNPREDICTABLE results.
Specifying the same register for RdLo and Rm has UNPREDICTABLE results.
Specifying the same register for RdHi and RdLo has UNPREDICTABLE results.

**Early termination:** If the multiplier implementation supports early termination, it must be implemented on the value of the Rs operand. The type of early termination used (signed or unsigned) is IMPLEMENTATION DEFINED.

**ARM Architecture Reference Manual**
ARM DUI 0100B

**Description**  The SMULL (Signed Multiply Long) instruction multiplies signed variables
to produce a 64-bit result in two general-purpose registers.

SMULL multiplies the signed value of register Rm with the signed value of register
Rs to produce a 64-bit result. The upper 32 bits of the result are stored in RdHi;
the lower 32 bits are stored in RdLo. The condition code flags are optionally
updated (based on the 64-bit result).

The instruction is only executed if the condition specified in the instruction matches
the condition code status. The conditions are defined in *3.3 The Condition Field* on
page 3-4.

| 31    28 | 27 26 25 24 23 22 21 | 20 | 19        16 | 15        12 | 11        8 | 7 6 5 4 | 3        0 |
|----------|----------------------|-----|--------------|--------------|-------------|---------|------------|
| cond     | 0  0  0  0  1  1  0   | S   | RdHi         | RdLo         | Rs          | 1 0 0 1 | Rm         |

**Operation**

```
if ConditionPassed(<cond>) then
    RdHi = (Rm * Rs)[63:32]
    RdLo = (Rm * Rs)[31:0]
    if S == 1 then
        N Flag = RdHi[31]
        Z Flag = if (RdHi == 0) and (RdLo == 0) then 1 else 0
        C Flag = UNPREDICTABLE
        V Flag = UNPREDICTABLE
```

**Exceptions**  None

**Qualifiers**  Condition Code
S updates condition code flags N,Z

**Notes**  **Use of R15:** Specifying R15 for register RdHi, RdLo, Rm or Rs has UNPREDICTABLE
results.

**Operand restriction:** Specifying the same register for RdHi and Rm has
UNPREDICTABLE results.
Specifying the same register for RdLo and Rm has UNPREDICTABLE results.
Specifying the same register for RdHi and RdLo has UNPREDICTABLE results.

**Early termination:** If the multiplier implementation supports early termination,
it must be implemented on the value of the Rs operand. The type of early
termination used (signed or unsigned) is IMPLEMENTATION DEFINED.

STC{<cond>}  p<cp_num>, CRd, <addressing_mode>

**Description**    The STC (Store Coprocessor) instruction is useful for storing coprocessor data to memory. The N bit could be used to distinguish between a single- and double-precision transfer for a floating-point store instruction.

STC stores data from the coprocessor specified by `<cp_num>` to the sequence of consecutive memory addresses calculated by `<addressing_mode>`. If no coprocessors indicate that they can execute the instruction, an UNDEFINED instruction exception is generated.

The instruction is only executed if the condition specified in the instruction matches the condition code status. The conditions are defined in *3.3 The Condition Field* on page 3-4.

| 31      28 | 27 | 26 | 25 | 24 | 23 | 22 | 21 | 20 | 19      16 | 15      12 | 11      8 | 7                    0 |
|------------|----|----|----|----|----|----|----|----|------------|------------|-----------|------------------------|
| cond | 1 | 1 | 0 | P | U | N | W | 0 | Rn | CRd | cp_num | 8_bit_word_offset |

### Operation

```
if ConditionPassed(<cond>) then
    <address> = <start_address>
    while (NotFinished(coprocessor[<cp_num>]))
        Memory[<address>,4] = value from Coprocessor[<cp_num>]
        <address> = <address> + 4
    assert <address> == <end_address>
```

**Operation**    Undefined Instruction; Data Abort

**Qualifiers**    Condition Code

**Notes**    **Addressing mode:** The P, U and W bits specify the `<addressing_mode>`. See *Addressing Mode 5* starting on page 3-123.

**The N bit:** This bit is coprocessor-dependent. It can be used to distinguish between two sizes of data to transfer.

**Register Rn:** Specifies the base register used by `<addressing_mode>`.

**Coprocessor fields:** Only instruction bits[31:23], bits[21:16] and bits[11:0] are ARM architecture-defined; the remaining fields (bit[22] and bits[15:12]) are only recommendations, for compatibility with ARM Development Systems.

**Data Abort:** If a data abort is signalled and `<addressing_mode>` uses pre-indexed or post-indexed addressing, the value left in Rn is IMPLEMENTATION DEFINED, but is either the original base register value or the updated base register value.

**Non-word-aligned addresses:** Store coprocessor register instructions ignore the least-significant two bits of `<address>` (the words are not rotated as for load word).

**Alignment:** If an implementation includes a System Control Coprocessor (see *Chapter 7*), and alignment checking is enabled, an address with bits[1:0] != 0b00 will cause an alignment exception.

**ARM Architecture Reference Manual**

ARM DUI 0100B

**Description**  The STM (Store Multiple) instruction is useful as a block store instruction (combined with load multiple it allows efficient block copy) and for stack operations, including procedure entry to save general-purpose registers and the return address, and for updating the stack pointer.

STM stores a non-empty subset (or possibly all) of the general-purpose registers to sequential memory locations. The registers are stored in sequence, the lowest-numbered register first, to the lowest memory address (<start_address>); the highest-numbered register last, to the highest memory address (<end_address>).

The instruction is only executed if the condition specified in the instruction matches the condition code status. The conditions are defined in *3.3 The Condition Field* on page 3-4.

| 31      28 | 27 26 25 | 24 | 23 | 22 | 21 | 20 | 19      16 | 15                                    0 |
|------------|----------|----|----|----|----|----|------------|-----------------------------------------|
| cond       | 1 0 0    | P  | U  | 0  | W  | 0  | Rn         | register list                           |

**Operation**
```
if ConditionPassed(<cond>) then
        <address> = <start_address>
        for i = 0 to 15
            if <register_list>[i] == 1
                    Memory[<address>,4] = Ri
                    <address> = <address> + 4
        assert <end_address> == <address> - 4
```

**Exceptions**  Data Abort

**Qualifiers**  Condition Code
! sets the W bit, causing base register update

**Notes**  **Addressing mode:** The P, U and W bits distinguish between the different types of addressing mode. See *Addressing Mode 4* starting on page 3-116.

**Register Rn:** Specifies the base register used by <addressing_mode>.

**Use of R15:** If register 15 if specified as the base register Rn, the result is UNPREDICTABLE. If register 15 is specified in <register_list>, the value stored is IMPLEMENTATION DEFINED.

**Operand restrictions:** If Rn is specified in <register_list>, and writeback is specified, the stored value of Rn is UNPREDICTABLE.

**Data Abort:** If a data abort is signalled and <addressing_mode> specifies writeback, the value left in Rn is IMPLEMENTATION DEFINED, but is either the original base register value or the updated base register value.

**Non-word-aligned addresses:** STM instructions ignore the least-significant two bits of <address> ( words are not rotated as for load word).

**Alignment:** If an implementation includes a System Control Coprocessor (See *Chapter 7*), and alignment checking is enabled, an address with bits[1:0] != 0b00 will cause an alignment exception.

**ARM Architecture Reference Manual**
ARM DUI 0100B

STM{<cond>}<addressing_mode>  Rn{!}, <registers>^

**Description**

The STM (Store Multiple) instruction is used to store the user mode registers when the processor is in a privileged mode (useful when performing process swaps).

This form of STM stores a subset (or possibly all) of the user mode general-purpose registers (which are also the system mode general-purpose registers) to sequential memory locations. The registers are stored in sequence, the lowest-numbered register first, to the lowest memory address (<start_addr>); the highest-numbered register last, to the highest memory address (<end_addr>).

The instruction is only executed if the condition specified in the instruction matches the condition code status. The conditions are defined in *3.3 The Condition Field* on page 3-4.

| 31      28 | 27 26 25 | 24 | 23 | 22 | 21 | 20 19      16 | 15                                    0 |
|:----------:|:--------:|:--:|:--:|:--:|:--:|:-------------:|:---------------------------------------:|
| cond       | 1  0  0  | P  | U  | 1  | 0  | 0   Rn        | register list                           |

**Operation**

```
if ConditionPassed(<cond>) then
        <address> = <start_addr>
        for i = 0 to 15
            if <register_list>[i] == 1
                    Memory[<address>,4] = Ri_usr
                    <address> = <address> + 4
        assert <end_addr> == <address> - 4
```

**Exceptions**    Data Abort

**Qualifiers**    Condition Code

**Notes**

**Addressing mode:** The P and W bits distinguish between the different types of addressing mode. See *Addressing Mode 4* starting on page 3-116.

**Banked registers:** This instruction must not be followed by an instruction which accesses banked registers (a following NOP is a good way to ensure this).

**Writeback:** Setting bit 21 (the W bit) has UNPREDICTABLE results.

**User and System mode:** This instruction is UNPREDICTABLE in user or system mode.

**Register Rn:** Specifies the base register used by <addressing_mode>.

**Use of R15:** If register 15 is specified as the base register Rn, the result is UNPREDICTABLE. If register 15 is specified in <register_list> the value stored is IMPLEMENTATION DEFINED.

**Base register mode:** The base register is read from the current processor mode registers, not the user mode registers.

**Data Abort:** If a data abort is signalled, the value left in Rn is the original base register value.

**Non-word-aligned addresses:** Load multiple instructions ignore the least-significant two bits of <address> (the words are not rotated as for load word).

**Alignment:** If an implementation includes a System Control Coprocessor (see *Chapter 7*), and alignment checking is enabled, an address with bits[1:0] != 0b00 causes an alignment exception.

**ARM Architecture Reference Manual**

ARM DUI 0100B

STR{<cond>}  Rd, <addressing_mode>

**Description**  Combined with a suitable addressing mode, the STR (Store register) instruction stores 32-bit data from a general purpose register into memory. Using the PC as the base register allows PC-relative addressing, to facilitate position-independent code.

STR stores a word from register Rd to the memory address calculated by `<addressing_mode>`.

The instruction is only executed if the condition specified in the instruction matches the condition code status. The conditions are defined in *3.3 The Condition Field* on page 3-4.

| 31 | 28 | 27 26 25 | 24 | 23 | 22 | 21 | 20 | 19        16 | 15      12 | 11                                     0 |
|----|----|----------|----|----|----|----|----|--------------|------------|------------------------------------------|
| cond | | 0 1 I | P | U | 0 | W | 0 | Rn | Rd | addressing mode specific |

**Operation**  
```
if ConditionPassed(<cond>) then
        Memory[<address>,4] = Rd
```

**Exceptions**  Data Abort

**Qualifiers**  Condition Code

**Notes**  **Addressing modes:** The I, P, U and W bits specify the type of `<addressing_mode>` (see *Addressing Mode 2* starting on page 3-98).

**Register Rn:** Specifies the base register used by <addressing_mode>.

**Use of R15:** If register 15 is specified for Rd, the value stored is IMPLEMENTATION DEFINED.

**Operand restrictions:** If `<addressing_mode>` uses pre-indexed or post-indexed addressing, and the same register is specified for Rd and Rn, the results are UNPREDICTABLE.

**Data Abort:** If a data abort is signalled and `<addressing_mode>` uses preindexed or post-indexed addressing, the value left in Rn is IMPLEMENTATION DEFINED, but is either the original base register value or the updated base register value (even if the same register is specified for Rd and Rn).

**Alignment:** If an implementation includes a System Control Coprocessor (See *Chapter 7*), and alignment checking is enabled, an address with bits[1:0] != 0b00 will cause an alignment exception.

**ARM Architecture Reference Manual**  
ARM DUI 0100B

STR{<cond>}B  Rd, <addressing_mode>

Load and store

Addressing mode 2

**Description**  Combined with a suitable addressing mode, the STRB (Store Register Byte) writes the least-significant byte of a general-purpose register to memory.

Using the PC as the base register allows PC-relative addressing, to facilitate position-independent code.

STRB stores a byte from the least-significant byte of register Rd to the memory address calculated by <addressing_mode>.

The instruction is only executed if the condition specified in the instruction matches the condition code status. The conditions are defined in *3.3 The Condition Field* on page 3-4.

| 31    28 | 27 26 25 | 24 | 23 | 22 | 21 | 20 | 19    16 | 15    12 | 11    0 |
|----------|----------|----|----|----|----|----|----------|----------|---------|
| cond | 0 1 | I | P | U | 1 | W | 0 | Rn | Rd | addressing mode specific |

**Operation**  if ConditionPassed(<cond>) then
                  Memory[<address>,1] = Rd[7:0]

**Exceptions**  Data Abort

**Qualifiers**  Condition Code

**Notes**  **Addressing modes:** The I, P, U and W bits specify the type of <addressing_mode> (see *Addressing Mode 2* starting on page 3-98).

**Register Rn:** Specifies the base register used by <addressing_mode>.

**Use of R15:** If register 15 is specified for Rd, the result is UNPREDICTABLE.

**Operand restrictions:** If <addressing_mode> uses pre-indexed or post-indexed addressing, and the same register is specified for Rd and Rn, the results are UNPREDICTABLE.

**Data Abort:** If a data abort is signalled and <addressing_mode> uses pre-indexed or post-indexed addressing, the value left in Rn is IMPLEMENTATION DEFINED, but is either the original base register value or the updated base register value (even if the same register is specified for Rd and Rn).

**ARM Architecture Reference Manual**
ARM DUI 0100B

**Description**

The STRBT (Store Register Byte with Translation) instruction can be used by a (privileged) exception handler that is emulating a memory access instruction which would normally execute in User Mode. The access is restricted as if it has User Mode privilege.

STRBT stores a byte from the least-significant byte of register Rd to the memory address calculated by <post_indexed_addressing_mode>. If the instruction is executed when the processor is in a privileged mode, the memory system is signalled to treat the access as if the processor were in user mode.

The instruction is only executed if the condition specified in the instruction matches the condition code status. The conditions are defined in *3.3 The Condition Field* on page 3-4.

| 31 | 28 | 27 | 26 | 25 | 24 | 23 | 22 | 21 | 20 | 19 | 16 | 15 | 12 | 11 | 0 |
|----|----|----|----|----|----|----|----|----|----|----|----|----|----|----|---|
| cond | | 0 | 1 | I | 0 | U | 1 | 1 | 0 | Rn | | Rd | | addressing mode specific | |

**Operation**

```
if ConditionPassed(<cond>) then
        Memory[<address>,1] = Rd[7:0]
```

**Exceptions**    Data Abort

**Qualifiers**    Condition Code

**Notes**

**Addressing modes:** The I, P, and U bits specify the type of <addressing_mode> (see *Addressing Mode 2* starting on page 3-98).

**Register Rn:** Specifies the base register used by <post_indexed_addressing_mode>.

**User mode:** If this instruction is executed in user mode, an ordinary user mode access is performed.

**Use of R15:** If register 15 is specified for Rd, the result is UNPREDICTABLE.

**Operand restrictions:** If the same register is specified for Rd and Rn, the results are UNPREDICTABLE.

**Data Abort:** If a data abort is signalled, the value left in Rn is IMPLEMENTATION DEFINED, but is either the original base register value or the updated base register value (even if the same register is specified for Rd and Rn).

**ARM Architecture Reference Manual**
ARM DUI 0100B

# STRH

STR{<cond>}H  Rd, <addressing_mode>

*Load and store*

*Addressing mode 3*

*Architecture v4 only*

**Description**   Combined with a suitable addressing mode, the STRH (Store Register Halfword) instruction allows 16-bit data from a general-purpose register to be stored to memory. Using the PC as the base register allows PC-relative addressing, to facilitate position-independent code.

STRH stores a halfword from the least-significant halfword of register Rd to the memory address calculated by <addressing_mode>. If the address is not halfword-aligned, the result is UNPREDICTABLE.

The instruction is only executed if the condition specified in the instruction matches the condition code status. The conditions are defined in *3.3 The Condition Field* on page 3-4.

| 31        28 | 27 | 26 | 25 | 24 | 23 | 22 | 21 | 20 | 19        16 | 15        12 | 11        8 | 7 | 6 | 5 | 4 | 3        0 |
|---|---|---|---|---|---|---|---|---|---|---|---|---|---|---|---|---|
| cond | 0 | 0 | 0 | P | U | I | W | 0 | Rn | Rd | addr_mode | 1 | 0 | 1 | 1 | addr_mode |

**Operation**
```
if ConditionPassed(<cond>) then
    if <address>[0] == 0
        <data> = Rd[15:0]
    else /* <address>[0] == 1 */
        <data> = UNPREDICTABLE
    Memory[<address>,2] = <data>
```

**Exceptions**   Data Abort

**Qualifiers**   Condition Code

**Notes**   **Addressing modes:** The I, P, U and W bits specify the type of <addressing_mode> (see *Addressing Mode 3* starting on page 3-109).

**The addr_mode bits:** These bits are addressing-mode specific.

**Register Rn:** Specifies the base register used by <addressing_mode>.

**Use of R15:** If register 15 is specified for Rd, the result is UNPREDICTABLE.

**Operand restrictions:** If <addressing_mode> uses pre-indexed or post-indexed addressing, and the same register is specified for Rd and Rn, the results are UNPREDICTABLE.

**Data Abort:** If a data abort is signalled and <addressing_mode> uses pre-indexed or post-indexed addressing, the value left in Rn is IMPLEMENTATION DEFINED, but is either the original base register value or the updated base register value (even if the same register is specified for Rd and Rn).

**Non-half-word aligned addresses:** If the store address is not halfword-aligned, the stored value is UNPREDICTABLE.

**Alignment:** If an implementation includes a System Control Coprocessor (see *Chapter 7*), and alignment checking is enabled, an address with bit[0] != 0 will cause an alignment exception.

**ARM Architecture Reference Manual**

ARM DUI 0100B

**Description**

The STRT (Store Register with Translation) instruction can be used by a (privileged) exception handler that is emulating a memory access instruction that would normally execute in User Mode. The access is restricted as if it has User Mode privilege.

STRT stores a word from register Rd to the memory address calculated by `<post_indexed_addressing_mode>`. If the instruction is executed when the processor is in a privileged mode, the memory system is signalled to treat the access as if the processor was in user mode.

The instruction is only executed if the condition specified in the instruction matches the condition code status. The conditions are defined in *3.3 The Condition Field* on page 3-4.

| 31 | 28 | 27 | 26 | 25 | 24 | 23 | 22 | 21 | 20 | 19 | 16 | 15 | 12 | 11 | 0 |
|---|---|---|---|---|---|---|---|---|---|---|---|---|---|---|---|
| cond | | 0 | 1 | I | 0 | U | 0 | 1 | 0 | Rn | | Rd | | addressing mode specific | |

**Operation**

```
if ConditionPassed(<cond>) then
       Memory[<address>,4] = Rd
```

**Exceptions**   Data Abort

**Qualifiers**   Condition Code

**Notes**

**Addressing modes:** The I, P, and U bits specify the type of `<addressing_mode>` (see *Addressing Mode 2* starting on page 3-98).

**Register Rn:** Specifies the base register used by `<post_indexed_addressing_mode>`.

**User mode:** If this instruction is executed in user mode, an ordinary user mode access is performed.

**Use of R15:** If register 15 is specified for Rd, the value stored is IMPLEMENTATION DEFINED.

**Operand restrictions:** If the same register is specified for Rd and Rn, the results are UNPREDICTABLE.

**Data Abort:** If a data abort is signalled, the value left in Rn is IMPLEMENTATION DEFINED, but is either the original base register value or the updated base register value (even if the same register is specified for Rd and Rn).

**Alignment:** If an implementation includes a System Control Coprocessor (See *Chapter 7*), and alignment checking is enabled, an address with bits[1:0] != 0b00 will cause an alignment exception.

**ARM Architecture Reference Manual**

ARM DUI 0100B

`SUB{<cond>}{S}  Rd, Rn, <shifter_operand>`

**Description**

The SUB (Subtract) instruction is used to subtract one value from another to produce a third. To decrement a register value (in Rx) use:

`SUB Rx, Rx, #1`

SUB subtracts the value of `<shifter_operand>` from the value of register Rn, and stores the result in the destination register Rd. The condition code flags are optionally updated (based on the result).

SUBS is useful as a loop counter decrement, as the loop branch can test the flags for the appropriate termination condition, without the need for a `CMP Rx, #0`.

Use `SUBS PC, LR, #4` to return from an interrupt.

The instruction is only executed if the condition specified in the instruction matches the condition code status. The conditions are defined in *3.3 The Condition Field* on page 3-4.

| 31      28 | 27 26 | 25 | 24 23 22 21 | 20 | 19      16 | 15      12 | 11                      0 |
|------------|-------|----|-------------|----|------------|------------|---------------------------|
| cond       | 0  0  | I  | 0  0  1  0  | S  | Rn         | Rd         | shifter_operand           |

**Operation**

```
if ConditionPassed(<cond>) then
        Rd = Rn - <shifter_operand>
        if S == 1 and Rd == R15 then
            CPSR = SPSR
        else if S == 1 then
            N Flag = Rd[31]
            Z Flag = if Rd == 0 then 1 else 0
            C Flag = NOT BorrowFrom(Rn - <shifter_operand>)
            V Flag = OverflowFrom (Rn - <shifter_operand>)
```

**Exceptions**

None

**Qualifiers**

Condition Code
S updates condition code flags N,Z,C and V

**Notes**

**Shifter operand:** The shifter operands for this instruction are given in *Addressing Mode 1* starting on page 3-84.

**Writing to R15:** When Rd is R15 and the S flag in the instruction is not set, the result of the operation is placed in the PC. When Rd is R15 and the S flag is set, the result of the operation is placed in the PC and the SPSR corresponding to the current mode is moved to the CPSR. This allows state changes which atomically restore both PC and CPSR. This form of the instruction is UNPREDICTABLE in User mode and System mode.

**The I bit:** Bit 25 is used to distinguish between the immediate and register forms of `<shifter_operand>`.

**ARM Architecture Reference Manual**

ARM DUI 0100B

SWI{<cond>}   <24_bit_immediate>

**Description**  The SWI instruction causes a SWI exception, see *2.5 Exceptions* on page 2-6.

The SWI instruction is used as an operating system service call. It can be used in two ways:

- to use the 24-bit immediate value to indicate the OS service that is required
- to ignore the 24-bit field and indicate the service required with a general-purpose register

A SWI exception is generated, which is handled by an operating system to provide the requested service.

The instruction is only executed if the condition specified in the instruction matches the condition code status. The conditions are defined in *3.3 The Condition Field* on page 3-4.

| 31      28 | 27 26 25 24 | 23                                    0 |
|------------|-------------|------------------------------------------|
| cond | 1  1  1  1 | 24_bit_immediate |

**Operation**
```
if ConditionPassed(<cond>) then
        R14_svc = address of SWI instruction + 4
        SPSR_svc = CPSR
        CPSR[5:0] = 0b010011; enter Supervisor mode
        CPSR[7] = 1; disable IRQ
        PC = 0x08
```

**Exceptions**   None

**Qualifiers**   Condition Code

---

**ARM Architecture Reference Manual**
ARM DUI 0100B

# SWP

`SWP{<cond>}   Rd, Rm, [Rn]`

*Semaphore*

*Not in architecture
v1 or v2*

**Description**   The SWP (Swap) instruction swaps a word between registers and memory.

SWP loads a word from the memory address given by the value of register Rn. The value of register Rm is then stored to the memory address given by the value of Rn, and the original loaded value is written to register Rd. If the same register is specified for Rd and Rn, this instruction swaps the value of the register and the value at the memory address.

The instruction is only executed if the condition specified in the instruction matches the condition code status. The conditions are defined in *3.3 The Condition Field* on page 3-4.

| 31    28 | 27 26 25 24 23 22 | 21 20 19    16 | 15    12 | 11    8 | 7 6 5 4 | 3    0 |
|----------|-------------------|----------------|----------|---------|---------|--------|
| cond | 0 0 0 1 0 0 | SBZ | Rn | Rd | SBZ | 1 0 0 1 | Rm |

**Operation**

```
if ConditionPassed(<cond>) then
      <temp> = Memory[Rn,4]
      Memory[Rn,4] = Rm
      Rd = <temp>
```

**Exceptions**   Data Abort

**Qualifiers**   Condition Code

**Notes**   **Non-word-aligned addresses:** If the address is not word-aligned, the loaded value is rotated right by 8 times the value of `<address>[1:0]`.

**Use of R15:** If register 15 is specified for Rd, Rn or Rm, the result is UNPREDICTABLE.

**Operand restrictions:** If the same register is specified as Rn and Rm, or Rn and Rd, the result is UNPREDICTABLE.

**Data Abort:** If a data abort is signalled on either the load access or the store access (or both), the loaded value is not written to Rd.

**Alignment:** If an implementation includes a System Control Coprocessor (see *Chapter 7*), and alignment checking is enabled, an address with bits[1:0] != 0b00 will cause an alignment exception.

**ARM Architecture Reference Manual**

ARM DUI 0100B

`SWP{<cond>}B   Rd, Rm, [Rn]`

**Description**    The SWPB (Swap Byte) instruction swaps a byte between registers and memory.

SWPB loads a byte from the memory address given by the value of register Rn. The value of the least-significant byte of register Rm is stored to the memory address given by Rn, and the original loaded value is zero-extended to a 32-bit word, and the word is written to register Rd. If the same register is specified for Rd and Rn, this instruction swaps the value of the least-significant byte of the register and the byte value at the memory address.

The instruction is only executed if the condition specified in the instruction matches the condition code status. The conditions are defined in *3.3 The Condition Field* on page 3-4.

| 31      28 | 27 26 25 24 23 22 | 21 20 19 | 16 15 | 12 11 | 8 7 6 5 4 3 | 0 |
|---|---|---|---|---|---|---|
| cond | 0 0 0 1 0 1 | SBZ | Rn | Rd | SBZ | 1 0 0 1 | Rm |

**Operation**
```
if ConditionPassed(<cond>) then
        <temp> = Memory[Rn,1]
        Memory[Rn,1] = Rm[7:0]
        Rd = <temp>
```

**Exceptions**    Data Abort

**Qualifiers**    Condition Code

**Notes**    **Use of R15:** If register 15 is specified for Rd, Rn or Rm, the result is UNPREDICTABLE.

**Operand restrictions:** If the same register is specified as Rn and Rm, Rn and or Rd, the result is UNPREDICTABLE.

**Data Abort:** If a data abort is signalled on either the load access or the store access (or both), the loaded value is not written to Rd.

`TEQ{<cond>}  Rn, <shifter_operand>`

**Description**

The TEQ (Test equivalence) instruction is used to test if two values are equal, without affecting the V flag (as CMP does). TEQ is a lso useful for testing if two values have the same sign.

The comparison is the Logical Exclusive OR of the two operands.

TEQ performs a comparison by logically Exclusive ORing the value of register Rn with the value of `<shofter_operand>`, and updates the condition code flags (based on the result).

The instruction is only executed if the condition specified in the instruction matches the condition code status. The conditions are defined in *3.3 The Condition Field* on page 3-4.

| 31 28 | 27 26 | 25 | 24 23 22 21 | 20 | 19 16 | 15 12 | 11 0 |
|---|---|---|---|---|---|---|---|
| cond | 0 0 | I | 1 0 0 1 | 1 | Rn | SBZ | shifter_operand |

**Operation**

```
if ConditionPassed(<cond>) then
        <alu_out> = Rn EOR <shifter_operand>
        N Flag = <alu_out>[31]
        Z Flag = if <alu_out> == 0 then 1 else 0
        C Flag = <shifter_carry_out>
        V Flag = unaffected
```

**Exceptions**    None

**Qualifiers**    Condition Code

**Notes**    **Shifter operand:** The shifter operands for this instruction are given in *Addressing Mode 1* starting on page 3-84.

**The I bit:** Bit 25 is used to distinguish between the immediate and register forms of `<shifter_operand>`.

**ARM Architecture Reference Manual**

ARM DUI 0100B

**Description**   The TST (Test) instruction is used to determine if many bits of a register are all clear, or if at least one bit of a register is set. The comparison is a logical AND of the two operands.

TST performs a comparison by logically ANDing the value of register Rn with the value of <shifter_operand>, and updates the condition code flags (based on the result).

The instruction is only executed if the condition specified in the instruction matches the condition code status. The conditions are defined in *3.3 The Condition Field* on page 3-4.

| 31 | 28 | 27 | 26 | 25 | 24 | 23 | 22 | 21 | 20 | 19 | 16 | 15 | 12 | 11 | 0 |
|----|----|----|----|----|----|----|----|----|----|----|----|----|----|----|---|
| cond | | 0 | 0 | I | 1 | 0 | 0 | 0 | 1 | Rn | | SBZ | | shifter_operand | |

**Operation**   
```
if ConditionPassed(<cond>) then
        <alu_out> = Rn AND <shifter_operand>
        N Flag = <alu_out>[31]
        Z Flag = if <alu_out> == 0 then 1 else 0
        C Flag = <shifter_carry_out>
        V Flag = unaffected
```

**Exceptions**   None

**Qualifiers**   Condition Code

**Notes**   **Shifter operand:** The shifter operands for this instruction are given in *Addressing Mode 1* starting on page 3-84.

**The I bit:** Bit 25 is used to distinguish between the immediate and register forms of <shifter_operand>.

**ARM Architecture Reference Manual**
ARM DUI 0100B

## ARM UMLAL

*Multiply*

*Architecture v3M and v4 only*

UMLAL{<cond>}{<S>}  RdLo, RdHi, Rm, Rs

**Description**

The UMLAL (Unsigned Multiply Accumulate Long) instruction multiplies unsigned variables to produce a 64-bit result, which is added to the 64-bit value in the two destination general-purpose registers.The result is written back to the two destination general-purpose registers.

UMLAL multiplies the unsigned value of register Rm with the unsigned value of register Rs to produce a 64-bit result. The lower 32 bits of the result are added to RdLo and stored in RdLo; the upper 32 bits and the carry from the addition to RdLo are added to RdHi and stored in RdHi. The condition code flags are optionally updated (based on the 64-bit result).

The instruction is only executed if the condition specified in the instruction matches the condition code status. The conditions are defined in *3.3 The Condition Field* on page 3-4.

| 31    28 | 27 | 26 | 25 | 24 | 23 | 22 | 21 | 20 | 19    16 | 15    12 | 11    8 | 7 | 6 | 5 | 4 | 3    0 |
|----------|----|----|----|----|----|----|----|----|----------|----------|---------|---|---|---|---|--------|
| cond | 0 | 0 | 0 | 0 | 1 | 0 | 1 | S | RdHi | RdLo | Rs | 1 | 0 | 0 | 1 | Rm |

### Operation

```
if ConditionPassed(<cond>) then
    RdLo = (Rm * Rs)[31:0] + RdLo
    RdHi = (Rm * Rs)[63:32] + RdHi + CarryFrom((Rm * Rs)[31:0] + RdLo)
    if S == 1 then
        N Flag = RdHi[31]
        Z Flag = if (RdHi == 0) and (RdLo == 0) then 1 else 0
        C Flag = UNPREDICTABLE
        V Flag = UNPREDICTABLE
```

**Exceptions**    None

**Qualifiers**    Condition Code
S updates condition code flags N and Z

**Notes**

**Use of R15:** Specifying R15 for register RdHi, RdLo, Rm or Rs has UNPREDICTABLE results.

**Operand restriction:** Specifying the same register for RdHi and Rm has UNPREDICTABLE results.
Specifying the same register for RdLo and Rm has UNPREDICTABLE results.
Specifying the same register for RdHi and RdLo has UNPREDICTABLE results.

**Early termination:** If the multiplier implementation supports early termination, it must be implemented on the value of the Rs operand. The type of early termination used (signed or unsigned) is IMPLEMENTATION DEFINED.

**ARM Architecture Reference Manual**

ARM DUI 0100B

**Description**  The UMULL (Unsigned Multiply Long) instruction multiplies unsigned variables to produce a 64-bit result in two general-purpose registers.

UMULL multiplies the unsigned value of register Rm with the unsigned value of register Rs to produce a 64-bit result. The upper 32 bits of the result are stored in RdHi; the lower 32 bits are stored in RdLo. The condition code flags are optionally updated (based on the 64-bit result).

The instruction is only executed if the condition specified in the instruction matches the condition code status. See *3.3 The Condition Field* on page 3-4.

| 31 | 28 | 27 26 25 24 23 22 21 | 20 | 19 | 16 15 | 12 11 | 8 7 6 5 4 | 3 | 0 |
|---|---|---|---|---|---|---|---|---|---|
| cond | | 0 0 0 0 1 0 0 | S | RdHi | RdLo | Rs | 1 0 0 1 | Rm | |

**Operation**

```
if ConditionPassed(<cond>) then
    RdHi = (Rm * Rs)[63:32]
    RdLo = (Rm * Rs)[31:0]
    if S == 1 then
        N Flag = RdHi[31]
        Z Flag = if (RdHi == 0) and (RdLo == 0) then 1 else 0
        C Flag = UNPREDICTABLE
        V Flag = UNPREDICTABLE
```

**Exceptions**  None

**Qualifiers**  Condition Code
S updates condition code flags N and Z

**Notes**  **Use of R15:** Specifying R15 for register RdHi, RdLo, Rm or Rs has UNPREDICTABLE results.

**Operand restriction:** Specifying the same register for RdHi and Rm has UNPREDICTABLE results.
Specifying the same register for RdLo and Rm has UNPREDICTABLE results.
Specifying the same register for RdHi and RdLo has UNPREDICTABLE results.

**Early termination:** If the multiplier implementation supports early termination, it must be implemented on the value of the Rs operand. The type of early termination used (signed or unsigned) is IMPLEMENTATION DEFINED.

**ARM Architecture Reference Manual**
ARM DUI 0100B

# ARM Addressing Modes

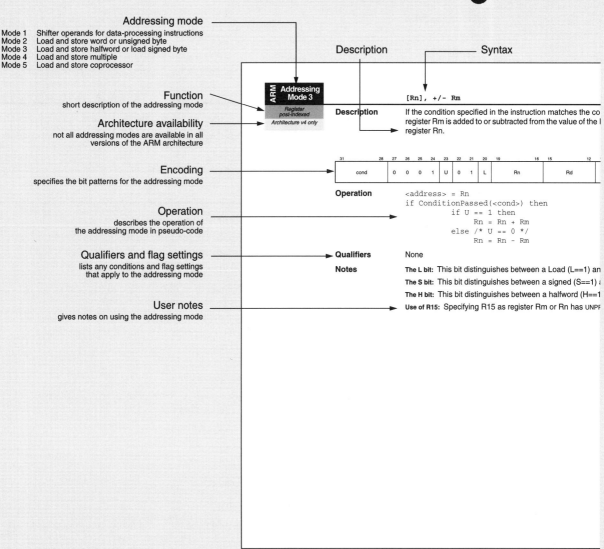

**Addressing mode**
Mode 1  Shifter operands for data-processing instructions
Mode 2  Load and store word or unsigned byte
Mode 3  Load and store halfword or load signed byte
Mode 4  Load and store multiple
Mode 5  Load and store coprocessor

**Function**
short description of the addressing mode

**Architecture availability**
not all addressing modes are available in all
versions of the ARM architecture

**Encoding**
specifies the bit patterns for the addressing mode

**Operation**
describes the operation of
the addressing mode in pseudo-code

**Qualifiers and flag settings**
lists any conditions and flag settings
that apply to the addressing mode

**User notes**
gives notes on using the addressing mode

**Description**          **Syntax**

ARM
Addressing
Mode 3
Register
post-indexed
Architecture v4 only

`[Rn], +/- Rm`

**Description**   If the condition specified in the instruction matches the co
register Rm is added to or subtracted from the value of the l
register Rn.

| 31 | 28 | 27 | 26 | 25 | 24 | 23 | 22 | 21 | 20 | 19 | 16 | 15 | 12 |
|----|----|----|----|----|----|----|----|----|----|----|----|----|----|
| cond | | 0 | 0 | 0 | 1 | U | 0 | 1 | L | Rn | | Rd | |

**Operation**
```
<address> = Rn
if ConditionPassed(<cond>) then
        if U == 1 then
              Rn = Rn + Rm
        else /* U == 0 */
              Rn = Rn - Rm
```

**Qualifiers**   None

**Notes**

**The L bit:** This bit distinguishes between a Load (L==1) an

**The S bit:** This bit distinguishes between a signed (S==1) a

**The H bit:** This bit distinguishes between a halfword (H==1

**Use of R15:** Specifying R15 as register Rm or Rn has UNPF

## 3.16 Data-processing Operands

```
<opcode>{<cond>}{S}{Rd}, {Rn}, <shifter_operand>
```

**32-bit immediate**

| 31    28 | 27 26 | 25 | 24    21 | 20 | 19    16 | 15    12 | 11     8 | 7              0 |
|----------|-------|----|----------|----|----------|----------|----------|------------------|
| cond     | 0  0  | 1  | opcode   | S  | Rn       | Rd       | rotate_imm | 8_bit_immediate |

**Immediate shifts**

| 31    28 | 27 26 | 25 | 24    21 | 20 | 19    16 | 15    12 | 11          7 | 6  5 | 4 | 3    0 |
|----------|-------|----|----------|----|----------|----------|---------------|------|---|--------|
| cond     | 0  0  | 0  | opcode   | S  | Rn       | Rd       | *shift imm*   | shift| 0 | Rm     |

**Register shifts**

| 31    28 | 27 26 | 25 | 24    21 | 20 | 19    16 | 15    12 | 11     8 | 7 | 6  5 | 4 | 3    0 |
|----------|-------|----|----------|----|----------|----------|----------|---|------|---|--------|
| cond     | 0  0  | 0  | opcode   | S  | Rn       | Rd       | Rs       | 0 | shift| 1 | Rm     |

**Description**

| | |
|---|---|
| <opcode> | Describes the operation of the instruction |
| S bit | Indicates that the instruction updates the condition codes. |
| Rd | Specifies the destination register |
| Rn | Specifies the first source operand register. |
| Bits[11:0] | The fields within bits[11:0] are collectively called a <shifter_operand>. This is described below. |
| Bit 25 | Is referred to as the I bit, and is used to distinguish between an immediate <shifter_operand> and a register-based <shifter_operand>. |

### 3.16.1 The shifter operand

As well as producing <shifter_operand>, the shifter produces a carry-out which some instructions write into the Carry Flag.

The shifter operand takes one of 3 basic formats:

* Immediate operand value
* Register operand value
* Shifted register operand value

**Format 1: Immediate operand value**

An immediate operand value is formed by rotating an 8-bit constant (in a 32-bit word) by an even number of bits (0,2,4,8...26,28,30). Thus, each instruction contains an 8-bit constant and a 4-bit rotate to be applied to that constant.

**ARM Architecture Reference Manual**

ARM DUI 0100B

Valid constants are:

```
0xff,0x104,0xff0,0xff00,0xff000,0xff000000,0xf000000f
```

Invalid constants are:

```
0x101,0x102,0xff1,0xff04,0xff003,0xffffffff,0xf000001f
```

For example:

```
MOV    R0, #0           ; Move zero to R0
ADD    R3, R3, #1       ; Add one to the value of register 3
CMP    R7, #1000        ; Compare value of R7 with 1000
BIC    R9, R8, #0xff00  ; Clear bits 8-15 of R8 and store in R9
```

### Format 2: Register operand value

A register operand value is simply the value of a register. The value of the register is used directly as the operand to the data-processing instruction.

For example:

```
MOV    R2, R0           ; Move the value of R0 to R2
ADD    R4, R3, R2       ; Add R2 to R3, store result in R4
CMP    R7, R8           ; Compare the value of R7 and R8
```

### Format 3: Shifted register operand value

A shifted register operand value is the value of a register, shifted (or rotated) before it is used as the data-processing operand. There are five types of shift:

| | |
|---|---|
| ASR | Arithmetic shift right |
| LSL | Logical shift left |
| LSR | Logical shift right |
| ROR | Rotate right |
| RRX | Rotate right with extend |

The number of bits to shift by is specified either as an immediate or as the value of the register.

For example:

```
MOV    R2, R0 LSL #2       ; Shift R0 left by 2, store in R2
                           ; (R2=R0x4)
ADD    R9, R5, R5 LSL #3   ; R9 = R5 + R5 x 8 or R9 = R5 x 9
RSB    R9, R5, R5 LSL #3   ; R9 = R5 x 8 - R5 or R9 = R5 x 7
SUB    R10, R9, R8 LSR #4  ; R10 = R9 - R8 / 16
MOV    R12, R4 ROR R3      ; R12 = R4 rotated right by value of R3
```

### The default shifter operand

The default register operand (register Rm specified with no shift) uses the form register shift left by immediate, with the immediate set to zero.

### 3.16.2 Shifter Operands

The 11 types of `<shifter_operand>` are described on the following pages:

**ARM Architecture Reference Manual**
ARM DUI 0100B

**Description**  The <shifter_operand> value is formed by rotating (to the right) an 8-bit immediate value to any even bit position in a 32-bit word. If the rotate immediate is zero, the carry-out from the shifter is the value of the C flag, otherwise, it is set to bit 31 of the value of <shifter_operand>.

This data-processing operand provides a constant (defined in the instruction) operand to a data-processing instruction.

| 31    28 | 27 26 25 | 24      21 | 20 | 19      16 | 15      12 | 11      8 | 7               0 |
|----------|----------|------------|----|------------|------------|-----------|-------------------|
| cond     | 0  0  1  | opcode     | S  | Rn         | Rd         | rotate_imm | 8_bit_immediate  |

**Operation**
```
<shifter_operand> = <8_bit_immediate> Rotate_Right
(<rotate_imm> * 2)
if <rotate_imm> == 0 then
        <shifter_carry_out> = C flag
else /* <rotate_imm> > 0 */
        <shifter_carry_out> = <shifter_operand>[31]
```

**Notes**  **Legitimate immediates:** Not all 32-bit immediates are legitimate; only those that can be formed by rotating an 8-bit immediate right by an even amount are valid 32-bit immediates for this format.

**Alternative assembly specification:** The 32-bit immediate can also be specified by:
```
#<8_bit_immediate>, <rotate_amount>
```
where:
```
<rotate_amount> = <rotate_imm> << 1
```

Rm

**Description**  This data-processing operand provides the value of a register directly.

This is an instruction operand produced by the value of register Rm. The carry-out from the shifter is the C flag.

| 31 | 28 | 27 26 25 | 24 | 21 | 20 | 19 | 16 | 15 | 12 | 11 10 9 8 7 | 6 5 4 | 3 | 0 |
|----|----|----------|----|----|----|----|----|----|----|-------------|-------|---|---|
| cond | | 0 0 0 | opcode | | S | Rn | | Rd | | 0 0 0 0 0 | 0 0 0 | Rm | |

**Operation**  `<shifter_operand> = Rm`
`<shifter_carry_out> = C Flag`

**Notes**  **Encoding:** This instruction is encoded as a Logical shift left by immediate (see page 3-89) with a shift of zero (`<shift_imm> == 0`).

**Use of R15:** If R15 is specified as register Rm or Rn, the value used is the address of the current instruction plus 8.

**ARM Architecture Reference Manual**
ARM DUI 0100B

**Description**    This data-processing operand is used to provide either the value of a register directly (lone register operand (see page 3-88), or the value of a register shifted left (multiplied by a constant power of two).

This instruction operand is produced by the value of register Rm, logically shifted left by an immediate value in the range 0 to 31. Zeros are inserted into the vacated bit positions. The carry-out from the shifter is the last bit shifted out, or the C flag if no shift is specified (lone register operand, see page 3-88).

| 31 | 28 | 27 26 25 | 24 | 21 | 20 | 19 | 16 | 15 | 12 | 11 | 7 | 6 5 4 | 3 | 0 |
|----|----|----------|------|----|----|----|----|----|----|-----------|---|-------|----|----|
| cond | | 0  0  0 | opcode | | S | Rn | | Rd | | shift_imm | | 0  0  0 | Rm | |

**Operation**
```
if <shift_imm> == 0 then /* Register Operand */
        <shifter_operand> = Rm
        <shifter_carry_out> = C Flag
else /* <shift_imm> > 0 */
        <shifter_operand> = Rm Logical_Shift_Left <shift_imm>
        <shifter_carry_out> = Rm[32 - <shift_imm>]
```

**Notes**    **Default shift:** If the value of <shift_imm> == 0, the operand may be written as just Rm, (see page 3-88).

**Use of R15:** If R15 is specified as register Rm or Rn, the value used is the address of the current instruction plus 8.

**ARM Architecture Reference Manual**
ARM DUI 0100B

Rm, LSL Rs

**Description**  This data-processing operand is used to provide the value of a register multiplied by a variable (in a register) power of two.

It is produced by the value of register Rm, logically shifted left by the value in the least-significant byte of register Rs. Zeros are inserted into the vacated bit positions.

| 31        28 | 27 26 25 | 24        21 | 20 | 19        16 | 15        12 | 11        8 | 7 6 5 4 | 3        0 |
|---|---|---|---|---|---|---|---|---|
| cond | 0  0  0 | opcode | S | Rn | Rd | Rs | 0  0  0  1 | Rm |

**Operation**
```
if Rs[7:0] == 0 then
        <shifter_operand> = Rm
        <shifter_carry_out> = C Flag
else if Rs[7:0] < 32 then
        <shifter_operand> = Rm Logical_Shift_Left Rs[7:0]
        <shifter_carry_out> = Rm[32 - Rs[7:0]]
else if Rs[7:0] == 32 then
        <shifter_operand> = 0
        <shifter_carry_out> = Rm[0]
else /* Rs[7:0] > 32 */
        <shifter_operand> = 0
        <shifter_carry_out> = 0
```

**Notes**  **Use of R15:** Specifying R15 as register Rm, register Rn or register Rs has UNPREDICTABLE results.

**ARM Architecture Reference Manual**
ARM DUI 0100B

**Description**    This data-processing operand is used to provide the unsigned value of a register shifted right (divided by a constant power of two).

It is produced by the value of register Rm logically shifted right by an immediate value in the range 1 to 32. Zeros are inserted into the vacated bit positions. A shift by 32 is encoded by `<shift_imm>` = 0.

| 31      28 | 27 26 25 | 24      21 | 20 | 19      16 | 15      12 | 11      7 | 6 5 4 | 3      0 |
|---|---|---|---|---|---|---|---|---|
| cond | 0 0 0 | opcode | S | Rn | Rd | shift_imm | 0 1 0 | Rm |

**Operation**
```
if <shift_imm> == 0 then
        <shifter_operand> = 0
        <shifter_carry_out> = Rm[31]
else /* <shift_imm> > 0 */
        <shifter_operand> = Rm Logical_Shift_Right <shift_imm>
        <shifter_carry_out> = Rm[<shift_imm> - 1]
```

**Notes**    **Use of R15:** If R15 is specified as register Rm or Rn, the value used is the address of the current instruction plus 8.

Rm, LSR Rs

*Logical shift right by register*

**Description**

This data-processing operand is used to provide the unsigned value of a register shifted right (divided by a variable power of two (in a register)).

It is produced by the value of register Rm logically shifted right by the value in the least-significant byte of register Rs. Zeros are inserted into the vacated bit positions.

| 31 28 | 27 26 25 | 24 21 | 20 | 19 16 | 15 12 | 11 8 | 7 6 5 4 | 3 0 |
|---|---|---|---|---|---|---|---|---|
| cond | 0 0 0 | opcode | S | Rn | Rd | Rs | 0 0 1 1 | Rm |

**Operation**

```
if Rs[7:0] == 0 then
        <shifter_operand> = Rm
        <shifter_carry_out> = C Flag
else if Rs[7:0] < 32 then
        <shifter_operand> = Rm Logical_Shift_Right Rs[7:0]
        <shifter_carry_out> = Rm[Rs[7:0] - 1]
else if Rs[7:0] == 32 then
        <shifter_operand> = 0
        <shifter_carry_out> = Rm[31]
else /* Rs[7:0] > 32 */
        <shifter_operand> = 0
        <shifter_carry_out> = 0
```

**Notes**

**Use of R15:** Specifying R15 as register Rm, register Rn or register Rs has UNPREDICTABLE results.

**ARM Architecture Reference Manual**
ARM DUI 0100B

**Description**  This data-processing operand is used to provide the signed value of a register arithmetically shifted right (divided by a constant power of two).

It is produced by the value of register Rm arithmetically shifted right by an immediate value in the range 1 to 32. The sign bit of Rm (Rm[31]) is inserted into the vacated bit positions. A shift by 32 is encoded by `<shift_imm>` = 0.

| 31    28 | 27 26 25 | 24    21 | 20 | 19    16 | 15    12 | 11    7 | 6 5 4 | 3    0 |
|----------|----------|----------|----|----------|----------|---------|-------|--------|
| cond     | 0  0  0  | opcode   | S  | Rn       | Rd       | shift_imm | 1 0 0 | Rm    |

**Operation**
```
if <shift_imm> == 0 then
      if Rm[31] == 0 then
          <shifter_operand> = 0
          <shifter_carry_out> = Rm[31]
      else /* Rm[31] == 1 */
          <shifter_operand> = 0xffffffff
          <shifter_carry_out> = Rm[31]
else /* <shift_imm> > 0 */
      <shifter_operand> = Rm Arithmetic_Shift_Right
<shift_imm>
      <shifter_carry_out> = Rm[<shift_imm> - 1]
```

**Notes**  **Use of R15:**  If R15 is specified as register Rm or Rn, the value used is the address of the current instruction plus 8.

`Rm, ASR Rs`

**Description** This data-processing operand is used to provide the signed value of a register arithmetically shifted right (divided by a variable power of two (in a register)).

It is produced by the value of register Rm arithmetically shifted right by the value in the least-significant byte of register Rs. The sign bit of Rm (Rm[31]) is inserted into the vacated bit positions.

| 31 28 | 27 26 25 | 24 21 | 20 | 19 16 | 15 12 | 11 8 | 7 6 5 4 | 3 0 |
|---|---|---|---|---|---|---|---|---|
| cond | 0 0 0 | opcode | S | Rn | Rd | Rs | 0 1 0 1 | Rm |

**Operation**
```
if Rs[7:0] == 0 then
        <shifter_operand> = Rm
        <shifter_carry_out> = C Flag
else if Rs[7:0] < 32 then
        <shifter_operand> = Rm Arithmetic_Shift_Right Rs[7:0]
        <shifter_carry_out> = Rm[Rs[7:0] - 1]
else /* Rs[7:0] >= 32 */
        if Rm[31] == 0 then
                <shifter_operand> = 0
                <shifter_carry_out> = Rm[31]
        else /* Rm[31] == 1 */
                <shifter_operand> = 0xffffffff
                <shifter_carry_out> = Rm[31]
```

**Notes** **Use of R15:** Specifying R15 as register Rm, register Rn or register Rs has UNPREDICTABLE results.

```
Rm, ROR #<shift_imm>
```

**Description**     This data-processing operand is used to provide the value of a register rotated by a constant value.

An instruction operand produced by the value of register Rm rotated right by an immediate value in the range 1 to 31. As bits are rotated off the right end, they are inserted into the vacated bit positions on the left.

When <shift_imm> = 0, a Rotate right with extend operation is performed; see page 3-97.

| 31 | 28 | 27 26 25 | 24 | 21 | 20 | 19 | 16 | 15 | 12 | 11 | 7 | 6 5 4 | 3 | 0 |
|---|---|---|---|---|---|---|---|---|---|---|---|---|---|---|
| cond | | 0 0 0 | opcode | | S | Rn | | Rd | | shift_imm | | 1 1 0 | Rm | |

**Operation**
```
if <shift_imm> == 0 then
        See Section , Rm, RRX, on page 3-97
else /* <shift_imm> > 0 */
        <shifter_operand> = Rm Rotate_Right <shift_imm>
        <shifter_carry_out> = Rm[<shift_imm> - 1]
```

**Notes**          **Use of R15:** If R15 is specified as register Rm or Rn, the value used is the address of the current instruction plus 8.

**ARM Architecture Reference Manual**
ARM DUI 0100B

`Rm, ROR Rs`

**Description** This data-processing operand is used to provide the value of a register rotated by a variable value (in a register).

It is produced by the value of register Rm rotated right by the value in the least-significant byte of register Rs. As bits are rotated off the right end, they are inserted into the vacated bit positions on the left.

| 31      28 | 27 26 25 | 24      21 | 20 | 19      16 | 15      12 | 11      8 | 7 6 5 4 | 3      0 |
|------------|----------|------------|----|------------|------------|-----------|---------|----------|
| cond | 0  0  0 | opcode | S | Rn | Rd | Rs | 0  1  1  1 | Rm |

**Operation**
```
if Rs[7:0] == 0 then
        <shifter_operand> = Rm
        <shifter_carry_out> = C Flag
else if Rs[4:0] == 0 then
        <shifter_operand> = Rm
        <shifter_carry_out> = Rm[31]
else /* Rs[4:0] > 0 */
        <shifter_operand> = Rm Rotate_Right Rs[4:0]
        <shifter_carry_out> = Rm[Rs[4:0] - 1]
```

**Notes** **Use of R15:** Specifying R15 as register Rm, register Rn or register Rs has UNPREDICTABLE results.

**Description**   This data-processing operand can be used to perform a 33-bit rotate right using the Carry Flag as the 33rd bit.

It is produced by the value of register Rm shifted right by one bit, with the Carry Flag replacing the vacated bit position.

| 31 | 28 | 27 26 25 | 24 | 21 | 20 | 19 | 16 | 15 | 12 | 11 10 9 8 7 6 5 4 | 3 | 0 |
|----|----|----------|--------|----|----|----|----|----|----|----|------|
| cond | | 0 0 0 | opcode | | S | Rn | | Rd | | 0 0 0 0 0 1 1 0 | Rm | |

**Operation**   ```
<shifter_operand> = (C Flag Logical_Shift_Left 31) OR
                    (Rm Logical_Shift_Right 1)
<shifter_carry_out> = Rm[0]
```

**Notes**   **Encoding:** The instruction encoding is in the space that would be used for `ROR #0`.

**Use of R15:** If R15 is specified as register Rm or Rn, the value used is the address of the current instruction plus 8.

**ADC instruction:** A rotate right with extend can be performed with an ADC instruction.

**ARM Architecture Reference Manual**
ARM DUI 0100B
```

## 3.17 Load and Store Word or Unsigned Byte Addressing Modes

There are nine addressing modes used to calculate the address for a load and store word or unsigned byte instruction. Each addressing mode is described in detail on the following pages.

Immediate offset                                                page 3-100

```
LDR|STR{<cond>}{B}  Rd, [Rn, #+/-<12_bit_offset>]
```

Register offset                                                 page 3-101

```
LDR|STR{<cond>}{B}  Rd, [Rn, +/-Rm]
```

Scaled register offsets                                         page 3-102

```
LDR|STR{<cond>}{B}  Rd, [Rn, +/-Rm, <shift> #<shift_imm>]
```

Immediate pre-indexed                                           page 3-103

```
LDR|STR{<cond>}{B}  Rd, [Rn, #+/-<12_bit_offset>]!
```

Register pre-indexed                                            page 3-104

```
LDR|STR{<cond>}{B}  Rd, [Rn, +/-Rm]!
```

Scaled register pre-indexed                                     page 3-105

```
LDR|STR{<cond>}{B}  Rd, [Rn, +/-Rm, <shift> #<shift_imm>]!
```

Immediate post-indexed                                          page 3-106

```
LDR|STR{<cond>}{B}{T}Rd, [Rn], #+/-<12_bit_offset>
```

Register post-indexed                                           page 3-107

```
LDR|STR{<cond>}{B}{T}Rd, [Rn], +/-Rm
```

Scaled register post-indexed                                    page 3-108

```
LDR|STR{<cond>}{B}{T}Rd, [Rn], +/-Rm, <shift> #<shift_imm>
```

**Immediate offset/index**

| 31    28 | 27 26 | 25 | 24 | 23 | 22 | 21 | 20 | 19    16 | 15    12 | 11                    0 |
|----------|-------|----|----|----|----|----|----|----------|----------|-------------------------|
| cond     | 0  1  | 0  | P  | U  | B  | W  | L  | Rn       | Rd       | 12_bit_offset           |

**Register offset/index**

| 31    28 | 27 26 | 25 | 24 | 23 | 22 | 21 | 20 | 19    16 | 15    12 | 11 10 9 8 7 6 5 4 | 3    0 |
|----------|-------|----|----|----|----|----|----|----------|----------|-------------------|--------|
| cond     | 0  1  | 1  | P  | U  | B  | W  | L  | Rn       | Rd       | 0 0 0 0 0 0 0 0   | Rm     |

**Scaled register offset/index**

| 31    28 | 27 26 | 25 | 24 | 23 | 22 | 21 | 20 | 19    16 | 15    12 | 11      7 | 6 5 | 4 | 3    0 |
|----------|-------|----|----|----|----|----|----|----------|----------|-----------|-----|---|--------|
| cond     | 0  1  | 1  | P  | U  | B  | W  | L  | Rn       | Rd       | shift_imm | shift | 0 | Rm   |

**ARM Architecture Reference Manual**
ARM DUI 0100B

**Notes**

**The P bit: Pre/Post indexing:**

Pre-indexing (P==1) indicates the offset is applied to the base register, and the result is used as the address.

Post-indexing (P==0) indicates the base register value is used for the address; the offset is then applied to the base register and written back to the base register.

**The U bit:** Indicates whether the offset is added to the base (U == 1) or subtracted from the base (U == 0).

**The B bit:** This bit distinguishes between an unsigned byte (B == 1) and a word (B == 0) access.

**The W bit:** This bit has two meanings:

| | |
|---|---|
| if P == 1 | if W == 1, the calculated address will be written back to the base register. (If W == 0, the base register will not be updated.) |
| if P == 0 | if W == 1, the current access is treated (by the protection and memory system) as a user mode access (if W == 0, a normal access is performed) |

**The L bit:** This bit distinguishes between a Load (L == 1) and a Store (L == 0).

`[Rn, #+/-<12_bit_offset>]`

**Description**
This addressing mode is useful for accessing structure (record) fields, and accessing parameters and locals variable in a stack frame. With an offset of zero, the address produced is the unaltered value of the base register Rn.

It calculates an address by adding or subtracting the value of an immediate offset to or from the value of the base register Rn.

| 31    28 | 27 | 26 | 25 | 24 | 23 | 22 | 21 | 20 | 19    16 | 15    12 | 11                    0 |
|----------|----|----|----|----|----|----|----|----|----------|----------|-------------------------|
| cond     | 0  | 1  | 0  | 1  | U  | B  | 0  | L  | Rn       | Rd       | 12_bit_offset           |

**Operation**
```
if U == 1 then
        <address> = Rn + <12_bit_offset>
else /* U == 0 */
        <address> = Rn - <12_bit_offset>
```

**Notes**
**The B bit:** This bit distinguishes between an unsigned byte (B==1) and a word (B==0) access.

**The L bit:** This bit distinguishes between a Load (L==1) and a Store (L==0) instruction.

**Use of R15:** If R15 is specified as register Rn, the value used is the address of the instruction plus 8.

**ARM Architecture Reference Manual**
ARM DUI 0100B

**Description**  This addressing mode is used for pointer + offset arithmetic, and accessing a single element of an array.

It calculates an address by adding or subtracting the value of the index register Rm to or from the value of the base register Rn.

| 31      28 | 27 | 26 | 25 | 24 | 23 | 22 | 21 | 20 | 19      16 | 15      12 | 11 | 10 | 9 | 8 | 7 | 6 | 5 | 4 | 3      0 |
|------------|----|----|----|----|----|----|----|----|------------|------------|----|----|---|---|---|---|---|---|----------|
| cond | 0 | 1 | 1 | 1 | U | B | 0 | L | Rn | Rd | 0 | 0 | 0 | 0 | 0 | 0 | 0 | 0 | Rm |

**Operation**

```
if U == 1 then
        <address> = Rn + Rm
else /* U == 0 */
        <address> = Rn - Rm
```

**Notes**  **Encoding:** This addressing mode is encoded as an LSL scaled register offset, scaled by zero.

**The B bit:** This bit distinguishes between an unsigned byte (B==1) and a word (B==0) access.

**The L bit:** This bit distinguishes between a Load (L==1) and a Store (L==0) instruction.

**Use of R15:** If R15 is specified as register Rn, the value used is the address of the instruction plus 8. Specifying R15 as register Rm has UNPREDICTABLE results.

```
[Rn, +/-Rm, LSL #<shift_imm>]
[Rn, +/-Rm, LSR #<shift_imm>]
[Rn, +/-Rm, ASR #<shift_imm>]
[Rn, +/-Rm, ROR #<shift_imm>]
[Rn, +/-Rm, RRX]
```

**Description**  These addressing modes are used for accessing a single element of an array of values larger than a byte.

They calculate an address by adding or subtracting the shifted or rotated value of the index register Rm to or from the value of the base register Rn.

| 31 28 | 27 26 25 24 | 23 | 22 | 21 | 20 | 19 16 | 15 12 | 11 7 | 6 5 | 4 | 3 0 |
|---|---|---|---|---|---|---|---|---|---|---|---|
| cond | 0 1 1 1 | U | B | 0 | L | Rn | Rd | shift_imm | shift | 0 | Rm |

**Operation**
```
case <shift> of
        00 /* LSL */
            <index> = Rm Logical_Shift_Left <shift_imm>
        01 /* LSR */
            <index> = Rm Logical_Shift_Right <shift_imm>
        10 /* ASR */
            <index> = Rm Arithmetic_Shift_Right <shift_imm>
        11 /* ROR or RRX */
            if <shift_imm> == 0 then /* RRX */
                <index> = (C Flag Logical_Shift_Left 31)
                                    OR (Rm Logical_Shift_Right 1)
            else /* ROR */
                <index> = Rm Rotate_Right <shift_imm>
    endcase
    if U == 1 then
        <address> = Rn + <index>
    else /* U == 0 */
        <address> = Rn - <index>
```

**Notes**  **The B bit:** This bit distinguishes between an unsigned byte (B==1) and a word (B==0) access.

**The L bit:** This bit distinguishes between a Load (L==1) and a Store (L==0) instruction.

**Use of R15:** If R15 is specified as register Rn, the value used is the address of the instruction plus 8. Specifying R15 as register Rm has UNPREDICTABLE results.

what do LSR, ASR #0 do?

**ARM Architecture Reference Manual**

ARM DUI 0100B

```
[Rn, #+/-<12_bit_offset>]!
```

**Description**  This addressing mode is used for pointer access to arrays with automatic update of the pointer value.

It calculates an address by adding or subtracting the value of an immediate offset to or from the value of the base register Rn.

If the condition specified in the instruction matches the condition code status, the calculated address is written back to the base register Rn. The conditions are defined in *3.3 The Condition Field* on page 3-4.

| 31      28 | 27 26 | 25 24 | 23 | 22 | 21 | 20 | 19      16 | 15      12 | 11                0 |
|------------|-------|-------|----|----|----|----|------------|------------|---------------------|
| cond       | 0  1  | 0  1  | U  | B  | 1  | L  | Rn         | Rd         | 12_bit_offset       |

**Operation**
```
if U == 1 then
        <address> = Rn + <12_bit_offset>
else /* if U == 0 */
        <address> = Rn - <12_bit_offset>
if ConditionPassed(<cond>) then
        Rn = <address>
```

**Notes**  **The B bit:** This bit distinguishes between an unsigned byte (B==1) and a word (B==0) access.

**The L bit:** This bit distinguishes between a Load (L==1) and a Store (L==0) instruction.

**Use of R15:** Specifying R15 as register Rn has UNPREDICTABLE results.

**ARM Architecture Reference Manual**
ARM DUI 0100B

[Rn, +/- Rm]!

*Register pre-indexed*

**Description**

This addressing mode calculates an address by adding or subtracting the value of an index register Rm to or from the value of the base register Rn.

If the condition specified in the instruction matches the condition code status, the calculated address is written back to the base register Rn. The conditions are defined in *3.3 The Condition Field* on page 3-4.

| 31 | 28 | 27 | 26 | 25 | 24 | 23 | 22 | 21 | 20 | 19 | 16 | 15 | 12 | 11 | 10 | 9 | 8 | 7 | 6 | 5 | 4 | 3 | 0 |
|----|----|----|----|----|----|----|----|----|----|----|----|----|----|----|----|----|----|----|----|----|----|----|----|
| cond | | 0 | 1 | 1 | 1 | U | B | 1 | L | Rn | | Rd | | 0 | 0 | 0 | 0 | 0 | 0 | 0 | 0 | Rm | |

**Operation**

```
if U == 1 then
        <address> = Rn + Rm
else /* U == 0 */
        <address> = Rn - Rm
if ConditionPassed(<cond>) then
        Rn = <address>
```

**Notes**

**The B bit:** This bit distinguishes between an unsigned byte (B==1) and a word (B==0) access.

**The L bit:** This bit distinguishes between a Load (L==1) and a Store (L==0) instruction.

**Use of R15:** Specifying R15 as register Rm or Rn has UNPREDICTABLE results.

**Operand Restrictions:** If the same register is specified for Rn and Rm, the result is UNPREDICTABLE.

**ARM Architecture Reference Manual**
ARM DUI 0100B

```
[Rn, +/-Rm, LSL #<shift_imm>]!
[Rn, +/-Rm, LSR #<shift_imm>]!
[Rn, +/-Rm, ASR #<shift_imm>]!
[Rn, +/-Rm, ROR #<shift_imm>]!
[Rn, +/-Rm, RRX]!
```

**Description**

These five addressing modes calculate an address by adding or subtracting the shifted or rotated value of the index register Rm to or from the value of the base register Rn.

If the condition specified in the instruction matches the condition code status, the calculated address is written back to the base register Rn. The conditions are defined in *3.3 The Condition Field* on page 3-4.

| 31 28 | 27 26 25 24 | 23 | 22 | 21 20 | 19 16 | 15 12 | 11 7 | 6 5 | 4 | 3 0 |
|---|---|---|---|---|---|---|---|---|---|---|
| cond | 0 1 1 1 | U | B | 1 L | Rn | Rd | shift_imm | shift | 0 | Rm |

**Operation**

```
case <shift> of
        00 /* LSL */
            <index> = Rm Logical_Shift_Left <shift_imm>
        01 /* LSR */
            <index> = Rm Logical_Shift_Right <shift_imm>
        10 /* ASR */
            <index> = Rm Arithmetic_Shift_Right <shift_imm>
        11 /* ROR or RRX */
            if <shift_imm> == 0 then /* RRX */
                <index> = (C Flag Logical_Shift_Left 31)
                OR (Rm Logical_Shift_Right 1)
            else /* ROR */
                <index> = Rm Rotate_Right <shift_imm>
endcase
if U == 1 then
        <address> = Rn + <index>
else /* U == 0 */
        <address> = Rn - <index>
if ConditionPassed(<cond>) then
        Rn = <address>
```

**Notes**

**The B bit:** This bit distinguishes between an unsigned byte (B==1) and a word (B==0) access.

**The L bit:** This bit distinguishes between a Load (L==1) and a Store (L==0) instruction.

**Use of R15:** Specifying R15 as register Rm or Rn has UNPREDICTABLE results.

**Operand Restrictions:** If the same register is specified for Rn and Rm, the result is UNPREDICTABLE.

**ARM Architecture Reference Manual**
ARM DUI 0100B

`[Rn], #+/-<12_bit_offset>`

**Description**  This addressing mode is used for pointer access to arrays with automatic update of the pointer value.

It calculates an address from the value of base register Rn.

If the condition specified in the instruction matches the condition code status, the value of the immediate offset added to or subtracted from the value of the base register Rn and written back to the base register Rn. The conditions are defined in *3.3 The Condition Field* on page 3-4

| 31  28 | 27 | 26 | 25 | 24 | 23 | 22 | 21 | 20 | 19  16 | 15  12 | 11  0 |
|---|---|---|---|---|---|---|---|---|---|---|---|
| cond | 0 | 1 | 0 | 0 | U | B | 0 | L | Rn | Rd | 12_bit_offset |

**Operation**
```
<address> = Rn
if ConditionPassed(<cond>) then
        if U == 1 then
              Rn = Rn + <12_bit_offset>
        else /* U == 0 */
              Rn = Rn - <12_bit_offset>
```

**Notes**  **The B bit:** This bit distinguishes between an unsigned byte (B==1) and a word (B==0) access.

**The L bit:** This bit distinguishes between a Load (L==1) and a Store (L==0) instruction.

**Use of R15:** Specifying R15 as register Rn has UNPREDICTABLE results.

**ARM Architecture Reference Manual**
ARM DUI 0100B

**Description**　This addressing mode calculates its address from the value of base register Rn.

If the condition specified in the instruction matches the condition code status, the value of the index register Rm is added to or subtracted from the value of the base register Rn and written back to the base register Rn. The conditions are defined in *3.3 The Condition Field* on page 3-4.

| 31 | 28 | 27 26 25 24 | 23 | 22 | 21 | 20 | 19　　16 | 15　　12 | 11 10 9 8 7 6 5 4 | 3　　0 |
|----|----|-------------|----|----|----|----|---------|---------|---------------------|--------|
| cond | | 0 1 1 0 | U | B | 0 | L | Rn | Rd | 0 0 0 0 0 0 0 0 | Rm |

**Operation**
```
<address> = Rn
if ConditionPassed(<cond>) then
        if U == 1 then
            Rn = Rn + Rm
        else /* U == 0 */
            Rn = Rn - Rm
```

**Notes**　**The B bit:** This bit distinguishes between an unsigned byte (B==1) and a word (B==0) access.

**The L bit:** This bit distinguishes between a Load (L==1) and a Store (L==0) instruction.

**Use of R15:** Specifying R15 as register Rn or Rm has UNPREDICTABLE results.

**Operand Restrictions:** If the same register is specified for Rn and Rm, the result is UNPREDICTABLE.

```
[Rn], +/-Rm, LSL #<shift_imm>
[Rn], +/-Rm, LSR #<shift_imm>
[Rn], +/-Rm, ASR #<shift_imm>
[Rn], +/-Rm, ROR #<shift_imm>
[Rn], +/-Rm, RRX
```

**Description**

If the condition specified in the instruction matches the condition code status, the shifted or rotated value of index register Rm is added to or subtracted from the value of the base register Rn and written back to the base register Rn. The conditions are defined in *3.3 The Condition Field* on page 3-4.

| 31    28 | 27 | 26 | 25 | 24 | 23 | 22 | 21 | 20 | 19    16 | 15    12 | 11    7 | 6 5 | 4 | 3    0 |
|----------|----|----|----|----|----|----|----|----|----------|----------|---------|-----|---|--------|
| cond     | 0  | 1  | 1  | 0  | U  | B  | 0  | L  | Rn       | Rd       | shift_imm | shift | 0 | Rm    |

**Operation**

```
<address> = Rn
case <shift> of
        00 /* LSL */
                <index> = Rm Logical_Shift_Left <shift_imm>
        01 /* LSR */
                <index> = Rm Logical_Shift_Right <shift_imm>
        10 /* ASR */
                <index> = Rm Arithmetic_Shift_Right <shift_imm>
        11 /* ROR or RRX */
                if <shift_imm> == 0 then /* RRX */
                    <index> = (C Flag Logical_Shift_Left 31)
                    OR (Rm Logical_Shift_Right 1)
                else /* ROR */
                    <index> = Rm Rotate_Right <shift_imm>
endcase
if ConditionPassed(<cond>) then
        if U == 1 then
            Rn = Rn + <index>
        else /* U == 0 */
            Rn = Rn - <index>
```

**Notes**

**The B bit:** This bit distinguishes between an unsigned byte (B==1) and a word (B==0) access.

**The L bit:** This bit distinguishes between a Load (L==1) and a Store (L==0) instruction.

**Use of R15:** Specifying R15 as register Rm or Rn has UNPREDICTABLE results.

**Operand Restrictions:** If the same register is specified for Rn and Rm, the result is UNPREDICTABLE.

**ARM Architecture Reference Manual**
ARM DUI 0100B

## 3.18 Load and Store Halfword or Load Signed Byte Addressing Modes

There are six addressing modes which are used to calculate the address for a load and store (signed or unsigned) halfword or load signed byte instructions.

1 Immediate offset                                                          page 3-110

    `LDR|STR{<cond>}H|SH|SB  Rd, [Rn, #+/-<8_bit_offset>]`

2 Register offset                                                           page 3-111

    `LDR|STR{<cond>}H|SH|SB  Rd, [Rn, +/-Rm]`

3 Immediate pre-indexed                                                     page 3-112

    `LDR|STR{<cond>}H|SH|SB  Rd, [Rn, #+/-<8_bit_offset>]!`

4 Register pre-indexed                                                      page 3-113

    `LDR|STR{<cond>}H|SH|SB  Rd, [Rn, +/-Rm]!`

5 Immediate post-indexed                                                    page 3-114

    `LDR|STR{<cond>}H|SH|SB  Rd, [Rn], #+/-<8_bit_offset>`

6 Register post-indexed                                                     page 3-115

    `LDR|STR{<cond>}H|SH|SB  Rd, [Rn], +/-Rm`

**Immediate offset/index**

| 31 | 28 | 27 | 26 | 25 | 24 | 23 | 22 | 21 | 20 | 19 | 16 | 15 | 12 | 11 | 8 | 7 | 6 | 5 | 4 | 3 | 0 |
|----|----|----|----|----|----|----|----|----|----|----|----|----|----|----|----|----|----|----|----|----|----|
| cond | | 0 | 0 | 0 | P | U | 1 | W | L | Rn | | Rd | | immedH | | 1 | S | H | 1 | ImmedL | |

**Register offset/index**

| 31 | 28 | 27 | 26 | 25 | 24 | 23 | 22 | 21 | 20 | 19 | 16 | 15 | 12 | 11 | 8 | 7 | 6 | 5 | 4 | 3 | 0 |
|----|----|----|----|----|----|----|----|----|----|----|----|----|----|----|----|----|----|----|----|----|----|
| cond | | 0 | 0 | 0 | P | U | 0 | W | L | Rn | | Rd | | SBZ | | 1 | S | H | 1 | Rm | |

**Notes**

**The P bit:** Pre-indexing (P==1) indicates the offset is applied to the base register, and the result is used as the address.
Post-indexing (P==0) indicates the base register value is used for the address; the offset is then applied to the base register and written back to the base register.

**The U bit:** Indicates whether the offset is added to the base (U==1) or subtracted from the base(U==0).

**The W bit:** If P is set, W indicates that the calculated address will be written back to the base register; if P is clear, the W bit must be clear or the instruction is UNPREDICTABLE.

**The L bit:** This bit distinguishes between a Load (L==1) and a Store (L==0) instruction.

**The S bit:** This bit distinguishes between a signed (S==1) and an unsigned (S==0) halfword access.

**The H bit:** This bit distinguishes between a halfword (H==1) and a signed byte (H==0) access.

**ARM Architecture Reference Manual**
ARM DUI 0100B

`[Rn, #+/-<8_bit_offset>]`

**Description**

This addressing mode is used for accessing structure (record) fields, and accessing parameters and locals variable in a stack frame. With an offset of zero, the address produced is the unaltered value of the base register Rn.

It calculates an address by adding or subtracting the value of an immediate offset to or from the value of the base register Rn.

| 31   28 | 27 26 25 24 | 23 | 22 21 20 | 19    16 | 15    12 | 11    8 | 7 | 6 | 5 | 4 | 3    0 |
|---------|-------------|-----|----------|----------|----------|---------|---|---|---|---|--------|
| cond | 0 0 0 1 | U | 1 0 L | Rn | Rd | ImmedH | 1 | S | H | 1 | ImmedL |

**Operation**

```
<8_bit_offset> = (<immedH> << 4) OR <immedL>
if U == 1 then
        <address> = Rn + <8_bit_offset>
else /* U == 0 */
        <address> = Rn - <8_bit_offset>
```

**Notes**

**The L bit:** This bit distinguishes between a Load (L==1) and a Store (L==0) instruction.

**The S bit:** This bit distinguishes between a signed (S==1) and an unsigned (S==0) halfword access.

**The H bit:** This bit distinguishes between a halfword (H==1) and a signed byte (H==0) access.

**Use of R15:** If R15 is specified as register Rn, the value used is the address of the instruction plus 8.

**ARM Architecture Reference Manual**
ARM DUI 0100B

**Description**　This addressing mode is useful for pointer + offset arithmetic, and for accessing a single element of an array.

It calculates an address by adding or subtracting the value of the index register Rm to or from the value of the base register Rn.

| 31 28 | 27 26 25 24 | 23 | 22 21 | 20 | 19 16 | 15 12 | 11 8 | 7 | 6 | 5 | 4 | 3 0 |
|---|---|---|---|---|---|---|---|---|---|---|---|---|
| cond | 0 0 0 1 | U | 0 0 | L | Rn | Rd | SBZ | 1 | S | H | 1 | Rm |

**Operation**
```
if U == 1 then
        <address> = Rn + Rm
else /* U == 0 */
        <address> = Rn - Rm
```

**Notes**　**The L bit:** This bit distinguishes between a Load (L==1) and a Store (L==0) instruction.

**The S bit:** This bit distinguishes between a signed (S==1) and an unsigned (S==0) halfword access.

**The H bit:** This bit distinguishes between a halfword (H==1) and a signed byte (H==0) access.

**Use of R15:** If R15 is specified as register Rn, the value used is the address of the instruction plus 8. Specifying R15 as register Rm has UNPREDICTABLE results.

`[Rn, #+/-<8_bit_offset>]!`

**Immediate pre-indexed**

**Architecture v4 only**

**Description**

This addressing mode gives pointer access to arrays, with automatic update of the pointer value.

It calculates an address by adding or subtracting the value of an immediate offset to or from the value of the base register `Rn`.

If the condition specified in the instruction matches the condition code status, the calculated address is written back to the base register Rn. The conditions are defined in *3.3 The Condition Field* on page 3-4.

| 31      28 | 27 26 25 24 | 23 | 22 21 | 20 | 19      16 | 15      12 | 11      8 | 7 | 6 | 5 | 4 | 3      0 |
|---|---|---|---|---|---|---|---|---|---|---|---|---|
| cond | 0 0 0 1 | U | 1 1 | L | Rn | Rd | immedH | 1 | S | H | 1 | ImmedL |

**Operation**

```
<8_bit_offset> = (<immedH> << 4) OR <immedL>
if U == 1 then
        <address> = Rn + <8_bit_offset>
else /* U == 0 */
        <address> = Rn - <8_bit_offset>
if ConditionPassed(<cond>) then
        Rn = <address>
```

**Notes**

**The L bit:** This bit distinguishes between a Load (L==1) and a Store (L==0) instruction.

**The S bit:** This bit distinguishes between a signed (S==1) and an unsigned (S==0) halfword access.

**The H bit:** This bit distinguishes between a halfword (H==1) and a signed byte (H==0) access.

**Use of R15:** Specifying R15 as register Rn has UNPREDICTABLE results.

`[Rn, +/- Rm]!`

**Description**    This addressing mode calculates an address by adding or subtracting the value of the index register Rm to or from the value of the base register Rn.

If the condition specified in the instruction matches the condition code status, the calculated address is written back to the base register Rn. The conditions are defined in *3.3 The Condition Field* on page 3-4.

| 31    28 | 27 26 25 24 | 23 | 22 21 | 20 | 19    16 | 15    12 | 11    8 | 7 | 6 | 5 | 4 | 3    0 |
|---|---|---|---|---|---|---|---|---|---|---|---|---|
| cond | 0 0 0 1 | U | 0 1 | L | Rn | Rd | SBZ | 1 | S | H | 1 | Rm |

**Operation**
```
if U == 1 then
        <address> = Rn + Rm
else /* U == 0 */
        <address> = Rn - Rm
if ConditionPassed(<cond>) then

        Rn = <address>
```

**Notes**    **The L bit:** This bit distinguishes between a Load (L==1) and a Store (L==0) instruction.

**The S bit:** This bit distinguishes between a signed (S==1) and an unsigned (S==0) halfword access.

**The H bit:** This bit distinguishes between a halfword (H==1) and a signed byte (H==0) access.

**Use of R15:** Specifying R15 as register Rm or Rn has UNPREDICTABLE results.

`[Rn], #+/-<8_bit_offset>`

**Description**

This addressing mode gives pointer access to arrays, with automatic update of the pointer value.

It calculates an address from the value of base register Rn.

If the condition specified in the instruction matches the condition code status, the value of the immediate offset is added to or subtracted from the value of the base register Rn and written back to the base register Rn. The conditions are defined in *3.3 The Condition Field* on page 3-4.

| 31      28 | 27 | 26 | 25 | 24 | 23 | 22 | 21 | 20 | 19      16 | 15      12 | 11      8 | 7 | 6 | 5 | 4 | 3      0 |
|---|---|---|---|---|---|---|---|---|---|---|---|---|---|---|---|---|
| cond | 0 | 0 | 0 | 0 | U | 1 | 0 | L | Rn | Rd | immedH | 1 | S | H | 1 | ImmedL |

**Operation**

```
<address> = Rn
<8_bit_offset> = (<immedH> << 4) OR <immedL>
if ConditionPassed(<cond>) then
        if U == 1 then
            Rn = Rn + <8_bit_offset>
        else /* U == 0 */
            Rn = Rn - <8_bit_offset>
```

**Notes**

**The L bit:** This bit distinguishes between a Load (L==1) and a Store (L==0) instruction.

**The S bit:** This bit distinguishes between a signed (S==1) and an unsigned (S==0) halfword access.

**The H bit:** This bit distinguishes between a halfword (H==1) and a signed byte (H==0) access.

**Use of R15:** Specifying R15 as register Rn has UNPREDICTABLE results.

**ARM Architecture Reference Manual**
ARM DUI 0100B

`[Rn], +/- Rm`

**Description**    If the condition specified in the instruction matches the condition code status, the value of the index register Rm is added to or subtracted from the value of the base register Rn and written back to the base register Rn. The conditions are defined in *3.3 The Condition Field* on page 3-4.

| 31 | 28 | 27 26 25 24 | 23 | 22 | 21 20 | 19 | 16 | 15 | 12 | 11 | 8 | 7 | 6 | 5 | 4 | 3 | 0 |
|---|---|---|---|---|---|---|---|---|---|---|---|---|---|---|---|---|---|
| cond | | 0 0 0 0 | U | 0 1 | L | Rn | | Rd | | SBZ | | 1 | S | H | 1 | Rm | |

**Operation**
```
<address> = Rn
if ConditionPassed(<cond>) then
        if U == 1 then
            Rn = Rn + Rm
        else /* U == 0 */
            Rn = Rn - Rm
```

**Notes**        **The L bit:** This bit distinguishes between a Load (L==1) and a Store (L==0) instruction.

   **The S bit:** This bit distinguishes between a signed (S==1) and an unsigned (S==0) halfword access.

   **The H bit:** This bit distinguishes between a halfword (H==1) and a signed byte (H==0) access.

   **Use of R15:** Specifying R15 as register Rm or Rn has UNPREDICTABLE results.

## 3.19   Load and Store Multiple Addressing Modes

Load Multiple instructions load a subset (possibly all) of the general-purpose registers from memory. Store Multiple instructions store a subset (possibly all) of the general purpose registers to memory. These instructions have a single instruction format.

Load and Store Multiple addressing modes produce a sequential range of addresses. The lowest-numbered register is stored at the lowest memory address and the highest-numbered register at the highest memory address.

There are four Load and Store Multiple addressing modes:

1   Increment After                                                          page 3-117

```
LDM|STM{<cond>}IA  Rn{!}, <registers>{^}
```

2   Increment Before                                                         page 3-118

```
LDM|STM{<cond>}IB  Rn{!}, <registers>{^}
```

3   Decrement After                                                          page 3-119

```
LDM|STM{<cond>}DA  Rn{!}, <registers>{^}
```

4   Decrement Before                                                         page 3-120

```
LDM|STM{<cond>}DB  Rn{!}, <registers>{^}
```

| 31    28 | 27 26 25 | 24 | 23 | 22 | 21 | 20 | 19    16 | 15    0 |
|----------|----------|----|----|----|----|----|----------|---------|
| cond | 1  0  0 | P | U | S | W | L | Rn | register list |

**Notes**

**The register list:** The `<register_list>` has 1 bit for each general-purpose register; bit 0 is for register zero, bit 15 is for register 15 (the PC). The `<register_list>` is specified in the instruction mnemonic using a comma-separated list of registers, surrounded by brackets. If no bits are set, the result is UNPREDICTABLE.

**The U bit:** Indicates that the transfer is made upwards (U==1) or downwards (U==0) from base register.

**The P bit:** Pre-indexing or post-indexing:

*Pre*   P==1   indicates that each address in the range is incremented (U==1) or decremented (U==0) before it is used to access memory.

*Post*   P==0   indicates that each address in the range is incremented (U==1) or decremented (U==0) after it is used to access memory.

**The W bit:** Indicates that the base register will be updated after the transfer. The base register is incremented (U==1) or decremented (U==0) by four times the number of registers in the register list.

**The S bit:** For LDMs that load the PC, the S bit indicates that the CPSR is loaded from the SPSR. For LDMs that do not load the PC and all STMs, the S bit indicates that when the processor is in a privileged mode, the user mode banked registers are transferred and not the registers of the current mode.

**The L bit:** Distinguishes between Load (L==1) and Store (L==0) instructions.

**ARM Architecture Reference Manual**

ARM DUI 0100B

`LDM|STM{<cond>}IA  Rd{!}, <registers>{^}`

**Description**   This addressing mode is for Load and Store multiple instructions, and forms a range of addresses.

The first address formed is the `<start_address>`, and is the value of the base register Rn. Subsequent addresses are formed by incrementing the previous address by four. One address is produced for each register that is specified in `<register_list>`.

The last address produced is the `<end_address>`; its value is four less than the sum of the value of the base register and four times the number of registers specified in `<register_list>`.

If the condition specified in the instruction matches the condition code status and the W bit is set, Rn is incremented by four times the numbers of registers in `<register_list>`. The conditions are defined in *3.3 The Condition Field* on page 3-4.

| 31    28 | 27 26 25 24 23 | 22 | 21 | 20 | 19    16 | 15    0 |
|----------|----------------|----|----|----|----------|---------|
| cond | 1 0 0 0 1 | S | W | L | Rn | register list |

**Operation**
```
<start_address> = Rn
<end_address> = Rn + (Number_Of_Set_Bits_In(<register_list>) * 4) - 4
if ConditionPassed(<cond>) and W == 1 then
        Rn = Rn + (Number_Of_Set_Bits_In(<register_list>) * 4)
```

**Qualifiers**   ! sets the W bit, causing base register update

^ is used to set the S bit, see below.

**Notes**   **The L bit:** This bit distinguishes between a load multiple and a store multiple.

**The S bit:** The action of the S bit is described in section *3.19 Load and Store Multiple Addressing Modes* on page 3-116.

**ARM Architecture Reference Manual**
ARM DUI 0100B

`LDM|STM{<cond>}IB  Rd{!}, <registers>{^}`

**Increment before**

**Description**   This addressing mode is for Load and Store multiple instructions, and forms a range of addresses.

The first address formed is the `<start_address>`, and is the value of the base register Rn plus four. Subsequent addresses are formed by incrementing the previous address by four. One address is produced for each register that is specified in `<register_list>`.

The last address produced is the `<end_address>`; its value is the sum of the value of the base register and four times the number of registers specified in `<register_list>`.

If the condition specified in the instruction matches the condition code status and the W bit is set, Rn is incremented by four times the numbers of registers in `<register_list>`. The conditions are defined in *3.3 The Condition Field* on page 3-4.

| 31    28 | 27 26 25 | 24 23 22 | 21 | 20 | 19    16 | 15    0 |
|----------|----------|----------|----|----|----------|---------|
| cond | 1 0 0 | 1 1 S | W | L | Rn | register list |

**Operation**
```
<start_address> = Rn + 4
<end_address> = Rn + (Number_Of_Set_Bits_In(<register_list>) * 4)
if ConditionPassed(<cond>) and W == 1 then
        Rn = Rn + (Number_Of_Set_Bits_In(<register_list>) * 4)
```

**Qualifiers**   ! sets the W bit, causing base register update

^ is used to set the S bit, see below.

**Notes**   **The L bit:** This bit distinguishes between a load multiple and a store multiple.

**The S bit:** The action of the S bit is described in section *3.19 Load and Store Multiple Addressing Modes* on page 3-116.

**ARM Architecture Reference Manual**
ARM DUI 0100B

**Description**  This addressing mode is for Load and Store multiple instructions, and forms a range of addresses.

The first address formed is the `<start_address>`, and is the value of the base register minus four times the number of registers specified in `<register_list>`, plus 4. Subsequent addresses are formed by incrementing the previous address by four. One address is produced for each register that is specified in `<register_list>`.

The last address produced is the `<end_address>`; its value is the value of base register Rn.

If the condition specified in the instruction matches the condition code status and the W bit is set, Rn is decremented by four times the numbers of registers in `<register_list>`. The conditions are defined in *3.3 The Condition Field* on page 3-4.

| 31 28 | 27 26 25 24 23 | 22 | 21 | 20 | 19 16 | 15 0 |
|---|---|---|---|---|---|---|
| cond | 1 0 0 0 0 | S | W | L | Rn | register list |

**Operation**

```
<start_address> = Rn - (Number_Of_Set_Bits_In(<register_list>) * 4) + 4
<end_address> = Rn
if ConditionPassed(<cond>) and W == 1 then
    Rn = Rn - (Number_Of_Set_Bits_In(<register_list>) * 4)
```

**Qualifiers**  ! sets the W bit, causing base register update

^ is used to set the S bit, see below.

**Notes**  **The L bit:** This bit distinguishes between a load multiple and a store multiple.

**The S bit:** The action of the S bit is described in section *3.19 Load and Store Multiple Addressing Modes* on page 3-116.

LDM|STM{<cond>}DB  Rd{!}, <registers>{^}

*Decrement before*

**Description**

This addressing mode is for Load and Store multiple instructions which form a range of addresses.

The first address formed is the <start_address>, and is the value of the base register minus four times the number of registers specified in <register_list>. Subsequent addresses are formed by incrementing the previous address by four. One address is produced for each register that is specified in <register_list>.

The last address produced is the <end_address>; its value is the value of base register Rn minus four.

If the condition specified in the instruction matches the condition code status and the W bit is set, Rn is decremented by four times the numbers of registers in <register_list>. The conditions are defined in *3.3 The Condition Field* on page 3-4.

| 31 | 28 | 27 | 26 | 25 | 24 | 23 | 22 | 21 | 20 | 19 | 16 | 15 | 0 |
|---|---|---|---|---|---|---|---|---|---|---|---|---|---|
| cond | | 1 | 0 | 0 | 1 | 0 | S | W | L | Rn | | register list | |

**Operation**

```
<start_address> = Rn - (Number_Of_Set_Bits_In(<register_list>) * 4)
<end_address> = Rn - 4
if ConditionPassed(<cond>) and W == 1 then
        Rn = Rn - (Number_Of_Set_Bits_In(<register_list>) * 4)
```

**Qualifiers**

! sets the W bit, causing base register update

^ is used to set the S bit, see below.

**Notes**

**The L bit:** This bit distinguishes between a load multiple and a store multiple.

**The S bit:** The action of the S bit is described in section *3.19 Load and Store Multiple Addressing Modes* on page 3-116.

**ARM Architecture Reference Manual**
ARM DUI 0100B

## 3.20 Load and Store Multiple Addressing Modes (Alternative names)

### 3.20.1 Block data transfer

The four addressing mode names given in *3.18 Load and Store Halfword or Load Signed Byte Addressing Modes* on page 3-109 (IA, IB, DA, DB) are most useful when a load and store multiple instruction is being used for block data transfer, as it is likely that the Load Multiple and Store Multiple will have the same addressing mode, so that the data is stored in the same way that it was loaded.

However, if Load Multiple and Store Multiple are being used to access a stack, the data will not be loaded with the same addressing mode that was used to store the data, because the load (pop) and store (push) operations must adjust the stack in opposite directions.

### 3.20.2 Stack operations

Load Multiple and Store Multiple addressing modes may be specified with an alternative syntax, which is more applicable to stack operations. Two attributes are used to describe the stack.

**Full or Empty**

| | |
|---|---|
| Full | is defined to have the stack pointer pointing to the last used (full) location in the stack |
| Empty | is defined to have the stack pointer pointing to the first unused (empty) location in the stack |

**Ascending or Descending**

| | |
|---|---|
| Descending | grows towards decreasing memory address (towards the bottom of memory) |
| Ascending | grows towards increasing memory address (towards the top of memory) |

This allows four types of stack to be defined:

1. Full Descending (FD)
2. Empty Descending (ED)
3. Full Ascending (FA)
4. Empty Ascending (EA)

*Table 3-1: LDM/STM addressing modes* shows the relationship between the four types of stack, the four types of addressing mode shown above, and the L, U and P bits in the instruction format:

**ARM Architecture Reference Manual**

ARM DUI 0100B

*Alternative names*

| Instruction | Addressing Mode | Stack Type | L bit | P Bit | U bit |
|---|---|---|---|---|---|
| LDM (Load) | IA (Increment After) | FD (Full Descending) | 1 | 0 | 1 |
| STM (Store) | IA (Increment After) | EA (Empty Ascending) | 0 | 0 | 1 |
| LDM (Load) | IB (Increment Before) | ED (Empty Descending) | 1 | 1 | 1 |
| STM (Store) | IB (Increment Before) | FA (Full Ascending) | 0 | 1 | 1 |
| LDM (Load) | DA (Decrement After) | FA (Full Ascending) | 1 | 0 | 0 |
| STM (Store) | DA (Decrement After) | ED (Empty Descending) | 0 | 0 | 0 |
| LDM (Load) | DB (Decrement Before) | EA (Empty Ascending) | 1 | 1 | 0 |
| STM (Store) | DB (Decrement Before) | FD (Full Descending) | 0 | 1 | 0 |

*Table 3-1: LDM/STM addressing modes*

**ARM Architecture Reference Manual**
ARM DUI 0100B

**ARM** POWERED

## 3.21   Load and Store Coprocessor Addressing Modes

There are three addressing modes which are used to calculate the address of a load or store coprocessor instruction:

1   Immediate offset                                     page 3-124

```
<opcode>{<cond>}{L}p<cp#>,CRd,[Rn,#+/-(<8_bit_offset>*4)]
```

2   Immediate pre-indexed                                page 3-125

```
<opcode>{<cond>}{L}p<cp#>,CRd,[Rn,#+/-(<8_bit_offset>*4)]!
```

3   Immediate post-indexed                               page 3-126

```
<opcode>{<cond>}{L}p<cp#>,CRd,[Rn],#+/-(<8_bit_offset>*4)
```

| 31 | 28 | 27 26 25 24 | 23 22 21 20 19 | | 16 15 | | 12 11 | | 8 7 | | 0 |
|---|---|---|---|---|---|---|---|---|---|---|---|
| cond | | 1 1 0 | P U N W L | Rn | | CRd | | cp# | | 8_bit_offset | |

**Notes**

**The P bit:**  Pre-indexing (P==1) or post-indexing (P==0):

(P==1)   indicates that the offset is added to the base register, and the result is used as the address.

(P==0)   indicates that the base register value is used for the address; the offset is then added to the base register and written back to the base register (because W will equal 1, see below).

**The U bit:**  Indicates that the offset is added to the base (U==1) or that the offset is subtracted from the base (U==0).

**The N bit:**  The meaning of this bit is coprocessor-dependent; its recommended use is to distinguish between different-sized values to be transferred.

**The W bit:**  This indicates that the calculated address will be written back to the base register. If P is 0, W must equal 1 or the result is UNPREDICTABLE.

**The L bit:**  Distinguishes between Load (L==1) and Store (L==0) instructions.

**ARM Architecture Reference Manual**

ARM DUI 0100B

`[Rn, #+/-(<8_bit_offset>*4)]`

*Immediate offset*

**Description**

This addressing mode produces a sequence of consecutive addresses.

The first address is calculated by adding or subtracting 4 times the value of an immediate offset to or from the value of the base register Rn. The subsequent addresses in the sequence are produced by incrementing the previous address by four until the coprocessor signals the end of the instruction. This allows a coprocessor to access data whose size is coprocessor-defined.

The coprocessor must not request a transfer of more than 16 words.

| 31    28 | 27 26 25 24 | 23 | 22 | 21 | 20 19    16 | 15    12 | 11    8 | 7    0 |
|----------|-------------|-----|-----|-----|-------------|----------|---------|--------|
| cond | 1 1 0 1 | U | N | 0 | L | Rn | CRd | cp_num | 8_bit_offset |

**Operation**

```
if ConditionPassed(<cond>) then
    if U == 1 then
        <address> = Rn + <8_bit_offset> * 4
    else /* U == 0 */
        <address> = Rn - <8_bit_offset> * 4
    <start_address> = <address>
    while (NotFinished(coprocessor[<cp_num>]))
        <address> = <address> + 4
    <end_address> = <address>
```

**Notes**

**The N bit:** This bit is coprocessor-dependent.

**The L bit:** Distinguishes between Load (L==1) and Store (L==0) instructions.

**Use of R15:** If R15 is specified as register Rn, the value used is the address of the instruction plus 8.

**ARM Architecture Reference Manual**
ARM DUI 0100B

`[Rn, #+/-(<8_bit_offset>*4)]!`

**Description**

This addressing mode produces a sequence of consecutive addresses.

The first address is calculated by adding or subtracting 4 times the value of an immediate offset to or from the value of the base register Rn. The first address is written back to the base register Rn. The subsequent addresses in the sequence are produced by incrementing the previous address by four until the coprocessor signals the end of the instruction. This allows a coprocessor to access data whose size is coprocessor-defined.

The coprocessor must not request a transfer of more than 16 words.

| 31 | 28 | 27 | 26 | 25 | 24 | 23 | 22 | 21 | 20 | 19 | 16 | 15 | 12 | 11 | 8 | 7 | 0 |
|---|---|---|---|---|---|---|---|---|---|---|---|---|---|---|---|---|---|
| cond | | 1 | 1 | 0 | 1 | U | N | 1 | L | Rn | | CRd | | cp_num | | 8_bit_offset | |

**Operation**

```
if ConditionPassed(<cond>) then
    if U == 1 then
        Rn = Rn + <8_bit_offset> * 4
    else /* U == 0 */
        Rn = Rn - <8_bit_offset> * 4
    <start_address> = Rn
    <address> = <start_address>
    while (NotFinished(coprocessor[<cp_num>]))
        <address> = <address> + 4
    <end_address> = <address>
```

**Notes**

**The N bit:** This bit is coprocessor-dependent.

**The L bit:** Distinguishes between Load (L==1) and Store (L==0) instructions.

**Use of R15:** If R15 is specified as register Rn, the value used is the address of the instruction plus 8.

**ARM Architecture Reference Manual**
ARM DUI 0100B

`[Rn], #+/-(<8_bit_offset>*4)`

**Description**

This addressing mode produces a sequence of consecutive addresses.

The first address is the value of the base register Rn. The subsequent addresses in the sequence are produced by incrementing the previous address by four until the coprocessor signals the end of the instruction. This allows a coprocessor to access data whose size is coprocessor-defined.

The base register Rn is updated by adding or subtracting 4 times the value of an immediate offset to or from the value of the base register Rn.

The coprocessor must not request a transfer of more than 16 words.

| 31    28 | 27 26 25 24 | 23 | 22 | 21 | 20 | 19      16 | 15    12 | 11     8 | 7           0 |
|----------|-------------|----|----|----|----|------------|----------|----------|---------------|
| cond     | 1 1 0 0     | U  | N  | 1  | L  | Rn         | CRd      | cp_num   | 8_bit_offset  |

**Operation**

```
if ConditionPassed(<cond>) then
    <start_address> = Rn
    if U == 1 then
        Rn = Rn + <8_bit_offset> * 4
    else /* U == 0 */
        Rn = Rn - <8_bit_offset> * 4
    <address> = <start_address>
    while (NotFinished(coprocessor[<cp_num>]))
        <address> = <address> + 4
    <end_address> = <address>
```

**Notes**

**The N bit:** This bit is coprocessor-dependent.

**The L bit:** Distinguishes between Load (L==1) and Store (L==0) instructions.

**Use of R15:** If R15 is specified as register Rn the value used is the address of the instruction plus 8.

**The W bit:** If bit 21 (the Writeback bit) is not set, the result is UNPREDICTABLE.

**ARM Architecture Reference Manual**
ARM DUI 0100B

**4**

# ARM Code Sequences

# 4

# ARM Code Sequences

The ARM instruction set is a powerful tool for generating high-performance microprocessor systems. Used to full extent, the ARM instruction set allows very compact and efficient algorithms to be coded. This chapter contains some sample routines that provide insight into the ARM instruction set.

# ARM Code Sequences

## 4.1 Arithmetic Instructions

The following code sequences illustrate some ways of using ARM's data-processing instructions.

### 4.1.1 Bit field manipulation

ARM shift and logical operations are very useful for bit manipulation:

```
; Extract 8 bits from the top of R2 and insert them into
; the bottom of R3
; R0 is a temporary value
    MOV  R0, R2, LSR #24   ; extract top bits from R2 into R0
    ORR  R3, R0, R3, LSL #8; shift up R3 and insert R0
```

### 4.1.2 Multiplication by constant

Combinations of shifts, add with shifts, and reverse subtract with shift can be used to perform multiplication by constant:

```
; multiplication of R0 by 2^n
    MOV  R0, R0, LSL #n          ; R0 = R0 << n
; multiplication of R0 by 2^n + 1
    ADD  R0, R0, R0, LSL #n      ; R0 = R0 + R0 << n
; multiplication of R0 by 2^n - 1
    RSB  R0, R0, LSL #n          ; R0 = R0 << n - R0

; R0 = R0 * 10 + R1
    ADD  R0, R0, R0, LSL #4      ; R0 = R0 * 5
    ADD  R0, R1, R0, LSL #1      ; R0 = R1 + R0 * 2

; R0 = R0 * 100 + R1, R2 is destroyed
    ADD  R2, R0, R0, LSL #3      ; R2 = R0 * 9
    ADD  R0, R2, R0, LSL #4      ; R0 = R2 + R0 * 16 (R0 = R0 * 25)
    ADD  R0, R1, R0, LSL #2      ; R0 = R1 + R0 * 4
```

**ARM Architecture Reference Manual**

ARM DDI 0100B

### 4.1.3 Multi-precision arithmetic

Arithmetic instructions allow efficient arithmetic on 64-bit or larger objects:

| | |
|---|---|
| Add, and Add with Carry | perform multi-precision addition |
| Subtract, and Subtract with Carry | perform subtraction |
| Compare | can be used for comparison |

```
; On entry :R0 and R1 hold a 64-bit number
; (R0 is least significant)
;           :R2 and R3 hold a second 64-bit number
; On exit  :R0 and R1 hold the 64-bit sum (or difference) of the 2 numbers
add64   ADDSR0, R0, R2      ; add lower halves and update Carry flag
        ADC R1, R1, R3      ; add the high halves and Carry flag

sub64   SUBSR0, R0, R2      ; subtract lower halves, update Carry
        SBC R1, R1, R3      ; subtract high halves and Carry

; This routine compares two 64-bit numbers
; On entry: As above
; On exit: N, Z, C and V flags updated correctly
cmp64   CMP R1, R3          ; compare high halves, if they are
        CMPEQR0, R2         ; equal, then compare lower halves
```

### 4.1.4 Swapping endianness

Swapping the order of bytes in a word (the endianness) can be performed in two ways.

1   The first method is best for single words:

```
; On entry:R0 holds the word to be swapped
; On exit: R0 holds the swapped word, R1 is destroyed
byteswap                    ; R0 = A , B , C , D
        EOR     R1, R0, R0, ROR #16  ; R1 = A^C,B^D,C^A,D^B
        BIC     R1, R1, #0xff0000    ; R1 = A^C, 0 ,C^A,D^B
        MOV     R0, R0, ROR #8       ; R0 = D , A , B , C
        EOR     R0, R0, R1, LSR #8   ; R0 = D , C , B , A
```

2   The second method is best for swapping the endianness of a large number of words:

```
; On entry: R0 holds the word to be swapped
; On exit : R0 holds the swapped word,
;         : R1, R2 and R3 are destroyed
byteswap        ; first the three instruction initialisation
        MOV     R2, #0xff            ; R2 = 0xff
        ORR     R2, R2, #0xff0000    ; R2 = 0x00ff00ff
        MOV     R3, R2, LSL #8       ; R3 = 0xff00ff00
        ; repeat the following code for each word to swap
                                     ; R0 = A   B   C   D
        AND     R1, R2, R0, ROR #24  ; R1 = 0   C   0   A
        AND     R0, R3, R0, ROR #8   ; R0 = D   0   B   0
        ORR     R0, R0, R1           ; R0 = D   C   B   A
```

## ARM Architecture Reference Manual
ARM DDI 0100B

# ARM Code Sequences

## 4.2 Branch Instructions

The following code sequences show some different ways of controlling the flow of execution in ARM code.

### 4.2.1 Procedure call and return

The BL (Branch and Link) instruction makes a procedure call by preserving the address of the instruction after the BL in R14 (the link register or LR), and then branching to the target address. Returning from a procedure is achieved by moving R14 to the PC:

```
        ....
        BL    function            ; call 'function'
        ....                       ; procedure returns to here
        ....
function                           ; function body
        ....
        ....
        MOV  PC, LR                ; Put R14 into PC to return
```

### 4.2.2 Conditional execution

Conditional execution allow if-then-else statements to be collapsed into sequences that do not require forward branches:

```
/* C code for Euclid's Greatest Common Divisor (GCD)*/
/* Returns the GCD of its two parameters */
int gcd(int a, int b)
{      while (a != b)
              if (a > b )
                      a = a - b ;
              else
                      b = b - a ;
       return a ;
}


; ARM assembler code for Euclid's Greatest Common Divisor
; On entry: R0 holds 'a', R1 holds 'b'
; On exit : R0 hold GCD of A and B
gcd    CMP    R0, R1    ; compare 'a' and 'b'
       SUBGT  R0, R0, R1; if (a>b) a=a-b (if a==b do nothing)
       SUBLT  R1, R1, R0; if (b>a) b=b-a (if a==b do nothing)
       BNE    gcd       ; if (a!=b) then keep going
       MOV    PC, LR    ; return to caller
```

**ARM Architecture Reference Manual**
ARM DDI 0100B

## 4.2.3 Conditional compare instructions

Compare instructions can be conditionally executed to implement more complicated expressions:

```
if (a==0 || b==1)
        c = d + e ;
CMP     R0, #0              ; compare a with 0
CMPNE   R1, #1              ; if a is not 0, compare b to 1
ADDEQ   R2, R3, R4          ; if either was true c = d + e
```

## 4.2.4 Loop variables

The subtract instruction can be used to both decrement a loop counter and set the condition codes to test for a zero.

```
    MOV         R0, #loopcount  ; initialise the loop counter
loop                            ; loop body
    ....
    ....
    SUBS        R0, R0, #1      ; subtract 1 from counter
                                ; set condition codes
    BNE         loop            ; if not zero, continue looping
    ....
```

## 4.2.5 Multi-way branch

A very simple multi-way branch can be implemented with a single instruction.
The following code dispatches the control of execution to any number of routines, with the restriction that the code to handle each case of the multi-way branch is the same size, and that size is a power of two bytes.

```
; Multi-way branch
; On entry: R0 holds the branch index
gcd     CMP     R0, #maxindex   ; (optional) checks the index is in range
        ADDLT   PC, PC, R0, LSL #RoutineSizeLog2
                                ; scale index by the log of the size of
                                ; each handler; add to the PC; jump there
        B       IndexOutOfRange; jump to the error handler
Index0Handler
        ....
        ....
Index1Handler
            ....
            ....
Index2Handler
            ....
            ....
Index3Handler
            ....
```

# ARM Code Sequences

## 4.3 Load and Store Instructions

Load and Store instructions are the best way to load or store a single word. They are also the only instructions that can load or store a byte or halfword.

### 4.3.1 Simple string compare

The following code performs a very simple string compare on two zero-terminated strings.

```
; String compare
; On entry : R0 points to the first string
;           : R1 points to the second string
;           : Call this code with a BL
; On exit  : R0 is < 0 if the first string is less than the second
;           : R0 is = 0 if the first string is equal to the second
;           : R0 is > 0 if the first string is greater than the second
;           : R1, R2 and R3 are destroyed

strcmp
    LDRB    R2, [R0] #1 ; get a byte from the first string
    LDRB    R3, [R1] #1; get a byte from the second string
    CMP     R2, #0; reached the end?
    BEQ     return; go to return code; calculate return value
    SUBS    R0, R2, R3; compare the two bytes
    BEQ     strcmp; if they are equal, keep looking
return
    MOV     PC, LR; put R14 (LR) into PC to return
```

Much faster implementations of this code are possible by loading a word of each string at a time and comparing all four bytes.

### 4.3.2 Linked lists

The following code searches for an element in a linked list. The linked list has two elements in each record; a single byte value and a pointer to the next record. A null next pointer indicates this is the last element in the list.

```
; Linked list search
; On entry: R0 holds a pointer to the first record in the list
; R1 holds the byte we are searching for
;           : Call this code with a BL
: On exit : R0 hold the address of the first record matched
: R2 is destroyed
or a null pointer if no match was found
llsearch
    CMP R0, #0              ; null pointer?
    LDRNEBR2, [R0]          ; load the byte value from this record
    CMPNER1, R2             ; compare with the lokked-for value
    LDRNER0, [R0, #4]       ; if not found, follow the link to the
    BNE llsearch            ; next record and then keep looking
    MOV PC, LR              ; return with pointer in R0
```

**ARM Architecture Reference Manual**
ARM DDI 0100B

## 4.3.3 Long branch

A load instruction can be used to generate a branch to anywhere in the 4Gbyte address space. By manually setting the value of the link register(R14), a subroutine call can be made to anywhere in the address space.

```
; Long branch (and link)
    ADD  LR, PC, #4      ; set the return address to be 8 byte
                         ; after the next instruction
    LDR  PC, [PC, #-4]   ; get the address from the next word
    DCD  function        ; store the address of the function
                         ; (DCD is an assembler directive)
                         ; return to here
```

This code uses the location after the load to hold the address of the function to call. In practice, this location can be accessing as long as it is within 4Kbytes of the load instruction. Notice also, that this code is position-independent except for the address of the function to call. Full position-independence can be achieved by storing the offset of the branch target after the load, and using an ADD instruction to add it to the PC.

## 4.3.4 Multi-way branches

The following code improves on the multi-way branch code shown above by using a table of addresses of functions to call.

```
; Multi-way branch
; On entry: R0 holds the branch index

    CMP   R0, #maxindex      ; (optional) checks the index is in range
    LDRLT PC, [PC, LSL #2]   ; convert the index to a word offset
                             ; do a look up in the table put the loaded
                             ; value into the PC and jump there
    B     IndexOutOfRange    ; jump to the error handler
    DCD   Handler0           ; DCD is an assembler directive to
    DCD   Handler1           ; store a word (in this case an
    DCD   Handler2           ; address in memory.
    DCD   Handler3
```

*, r0,*

# ARM Code Sequences

## 4.4    Load and Store Multiple Instructions

Load and Store Multiple instructions are the most efficient way to manipulate blocks of data.

### 4.4.1    Simple block copy

This code performs very simple block copy, 48 bytes at a time, and will approach the maximum throughput for a particular machine. The source and destination must be word-aligned, and objects with less than 48 bytes must be handled separately.

```
; Simple block copy function
; R12 points to the start of the source block
; R13 points to the start of the destination block
; R14 points to the end of the source block
loop    LDMIA   R12!, (R0-R11)   ; load 48 bytes
        STMIA   R13!, {R0-R11}   ; store 48 bytes
        CMP     R12, R14         ; reached the end yet?
        BLT     loop             ; branch to the top of the loop
```

### 4.4.2    Procedure entry and exit

This code uses load and store multiple to preserve and restore the processor state during a procedure. The code assumes the register r0 to r3 are argument registers, preserved by the caller of the function, and therefore do not need to be preserved. R13 is also assumed to point to a full descending stack.

```
function
    STMFD   R13!, {R4 - R12, R14}; preserve all the local registers
                                 ; and the return address, and
                                 ; update the stack pointer.

; function body

    LDMFD   R13!, {R4 - R12, PC} ; restore the local register, load
                                 ; the PC from the saved return
                                 ; update the stack pointer.
```

Notice that this code restores all saved registers, updates the stack pointer, and returns the caller (by loading the PC value) all in with single instruction. This allows very efficient conditional return for exceptional cases from a procedure (by checking the condition with a compare instruction and the conditionally executing the load multiple).

**ARM Architecture Reference Manual**
ARM DDI 0100B

## 4.5    Semaphore Instructions

This code controls the entry and exit from a critical section of code. The semaphore instructions do not provide a compare and conditional write facility; this must be done explicitly. The following code achieves this by using a semaphore value to indicate that the lock is being inspected.

The code below causes the calling process to busy-wait until the lock is free; to ensure progress, three OS calls need to be made (one before each loop branch) to sleep the process if the lock cannot be accessed.

```
        ; Critical section entry and exit
        ; The code uses a process ID to identify the lock owner
        ; An ID of zero indicates the lock is free
        ; An ID of -1 indicates the lock is being inspected
        ; On entry:  R0 holds the address of the semaphore
        ;            R1 holds the ID of the process requesting the lock

        MVN    R2, #0            ; load the 'looking' value (-1) in R2
spinin  SWP    R3, R2, [R0]      ; look at the lock, and lock others out
        CMN    R3, #1            ; anyone else trying to look?
        : conditional OS call to sleep process
        BEQ    spinin            ; yes, so wait our turn
        CMP    R3, #0            ; no-one looking, is the lock free?
        STRNE R3, [R0]           ; no, then restore the previous owner
        : conditional OS call to sleep process
        BNE    spinin            ; and wait again
        STR    R1, [R0]          ; otherwise grab the lock
        .....
        Insert critical code here
        .....
spinout SWP    R3, R2, [R0]      ; look at the lock, and lock others out
        CMN    R3, #1            ; anyone else trying to look ?
        : conditional OS call to sleep process
        BEQ    spinout           ; yes, so wait our turn
        CMP    R3, R1            ; check we own it
        BNE    CorruptSemaphore ; we should have been the owner!
        MOV    R2, #0            ; load the 'free' value
        STR    R2, [R0]          ; and open the lock
```

# ARM Code Sequences

## 4.6 Other Code Examples

The following sequences illustrate some other applications of ARM assembly language.

### 4.6.1 Software Interrupt dispatch

This code segment dispatches software interrupts (SWIs) to individual handlers.
The SWI instruction has a 24-bit field that can be used to encode specific SWI functions.

```
STMFD   SP!, {R12}              ; save some registers
LDR     R12, [R14, #-4]         ; load the SWI instruction
BIC     R12, R12, #0xff000000   ; preserve the SWI number
CMP     R12, #MaximumSWI        ; check the SWI number is in range, if so
LDRLE   PC, [PC, R12, LSL #2]   ; branch through a table to the handler
B       UnkownSWI               ; this SWI number is not supported

DCD     SWI0Handler             ; address of handler for SWI 0
DCD     SWI1Handler             ; address of handler for SWI 1
DCD     SWI2Handler             ; address of handler for SWI 2
        ....
```

### 4.6.2 Single-channel DMA transfer

The following code is an interrupt handler to perform interrupt driven IO to memory transfers (soft DMA). The code is especially useful as a FIQ handler, as it uses the banked FIQ registers to maintain state between interrupts. Therefore this code is best situated at location 0x1c.

R8          points to the base address of the IO device that data is read from

IOData      is the offset from the base address to the 32-bit data register that is read. Reading this register disables the interrupt

R9          points to the memory location where data is being transferred

R10         points to the last address to transfer to

The entire sequence to handle a normal transfer is just 4 instructions; code situated after the conditional return is used to signal that the transfer is complete.

```
LDR     r11, [r8, #IOData]     ; load port data from the IO device
STR     r11, [r9], #4          ; store it to memory: update the pointer
CMP     r9, r10                ; reached the end?
SUBLTS  pc, lr, #4             ; no, so return
; Insert transfer complete code here
```

Of course, byte transfers can be made by replacing the load instructions with load byte instructions, and transfers from memory to and IO device are made by swapping the addressing modes between the load instruction and the store instruction.

**ARM Architecture Reference Manual**
ARM DDI 0100B

## 4.6.3　Dual-channel DMA transfer

This code is similar to the example in *4.6.2 Single-channel DMA transfer* on page 4-10, except that it handles two channels (which may be the input and output side of the same channel). Again this code is especially useful as a FIQ handler, as it uses the banked FIQ registers to maintain state between interrupts. Therefore, this code is best situated at location 0x1c.

The entire sequence to handle a normal transfer is just 9 instructions; code situated after the conditional return is used to signal that the transfer is complete.

```
LDR     r13, [r8, #IOStat]    ; load status register to find ....
TST     r13, #IOPort1Active   ; .... which port caused the interrupt?
LDREQ   r13, [r8, #IOPort1]   ; load port 1 data
LDRNE   r13, [r8, #IOPort2]   ; load port 2 data
STREQ   r13, [r9], #4         ; store to buffer 1
STRNE   r13, [r10], #4        ; store to buffer 2
CMP     r9, r11               ; reached the end?
CMPNE   r10, r12              ; on either channel?
SUBNES  pc, lr, #4            ; return
; Insert transfer complete code here
```

where:

| | |
|---|---|
| R8 | points to the base address of the IO device that data is read from |
| IOStat | is the offset from the base address to a register indicating which of two ports caused the interrupt |
| IOPort1Active | is a bit mask indicating if the first port caused the interrupt (otherwise it is assumed that the second port caused the interrupt) |
| IOPort, IOPort2 | are offsets to the two data registers to be read. Reading a data register disables the interrupt for that port |
| R9 | points to the memory location that data from the first port is being transferred to |
| R10 | points to the memory location that data from the second port is being transferred to |
| R11 and R12 | point to the last address to transfer to (R11 for the first port, R12 for the second) |

Again, byte transfers can be made by replacing the load instructions with load byte instructions, and transfers from memory to and IO device are made by swapping the addressing modes between the conditional load instructions and the conditional store instructions.

# ARM Code Sequences

## 4.6.4 Interrupt prioritisation

This code dispatches up to 32 interrupt source to their appropriate handler routines. This code is intended to use the normal interrupt vector, and so should be branched to from location 0x18.

External hardware is used to prioritise the interrupt and present the number of the highest-priority active interrupt in an IO register.

IntBase     holds the base address of the interrupt handler

IntLevel     holds the offset (from IntBase) of the register containing the highest-priority active interrupt

R13     is assumed to point to a small (60 byte) full descending stack

Interrupts are enabled after 10 instructions (including the branch to this code)

```
        ; first save the critical state
        SUB   r14, r14, #4         ; adjust return address before saving it
        STMFD r13!, {r12, r14}     ; stack return address and working register
        MRS   r12, SPSR            ; get the SPSR ...
        STMFD r13!, {r12}          ; ... and stack that too
        ; now get the priority level of the highest priority active interrupt
        MOV   r12, #IntBase        ; get interrupt controller's base address
        LDR   r12, [r12, #IntLevel]; get the interrupt level (0 to 31)
        ; now read-modify-write the CPSR to enable interrupts
        MRS   r14, CPSR            ; read the status register
        BIC   r14, r14, #0x40      ; clear the F bit (use 0x80 for the I bit)
        MSR   CPSR, r14            ; write it back to re-enable interrupts
        ; jump to the correct handler
        LDR   PC, [PC, r12, LSL #2]; and jump to the correct handler
                                   ; PC base address points to this
                                   ; instruction + 8
        NOP                        ; pad so the PC indexes this table
        ; table of handler start addresses
        DCD   Priority0Handler
        DCD   Priority1Handler ........
Priority0Handler
        STMFD r13!, {r0 - r11}     ; save working registers
        ; insert handler code here
        ........
        LDMFD r13!, {r0 - r12}     ; restore the working registers and the
SPSR
        ORR   r14, r14, #0x40      ; set the F bit (use 0x80 for the I bit)
        MSR   CPSR, r14            ; write it back to disable interrupts
        MSR   SPSR, r12            ; stick the SPSR back
        LDMFD r13!, {r12, PC}^     ; restore last working register and return
Priority1Handler
        ........
```

**ARM Architecture Reference Manual**
ARM DDI 0100B

## 4.6.5 Context switch

This code performs a context switch on the user mode process. The code is based around a list of pointers to process control blocks (PCBs) of processes that are ready to run.

The pointer to the PCB of the next process to run is pointed to by R12, and the end of the list has a zero pointer.

R13 is a pointer to the PCB, and is preserved between timeslices (so that on entry R13 points to the PCB of the currently running process).

The code assumes the layout of the PCBs, as shown in *Figure 4-1: PCB layout.*

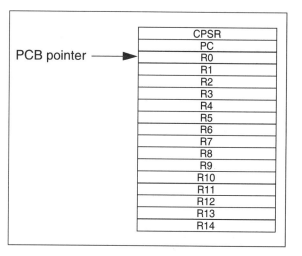

*Figure 4-1: PCB layout*

```
STMIA     r13, {r0 - r14}^     ; dump user registers above r13
MSR       r0, SPSR             ; pick up the user status
STMDB     r13, {r0, r14}       ; and dump with return address below
LDR       r13, [r12], #4       ; load next process info pointer
CMP       r13, #0              ; if it is zero, it is invalid
LDMNEDB   r13, {r0, r14}       ; pick up status and return address
MRSNE     SPSR, r0             ; restore the status
LDMNEIA   r13, {r0 - r14}^     ; get the rest of the registers
MOVNES    pc, r14              ; and return and restore CPSR
; insert "no next process code" here
```

# 5

# The 26-bit Architectures

# 5 The 26-bit Architectures

This chapter describes the differences between 32-bit and 26-bit architectures.

# The 26-bit Architectures

## 5.1    Introduction

ARM architecture versions 1, 2 and 2a are earlier versions of the ARM architecture which implement only a 26-bit address bus, and are known as 26-bit architectures. ARM architecture versions 3 and 4 implement a 32-bit address space (and are known as 32-bit architectures). For backwards compatibility, they also implement the 26-bit address space (except Version 3G). Implementation of a backwards-compatible 26-bit address space on ARM architecture version 4 is optional.

There are several differences between the 26-bit and the 32-bit architectures:

| | |
|---|---|
| Program counter | The 26-bit architectures implement only a 24-bit program counter in register 15, which allows 64Mbytes of program space. The 32-bit architectures have a 30-bit program counter in register 15 which allows 4Gbytes of program space on 32-bit architectures. |
| Processor modes | Only four processor modes are supported on 26-bit architectures: |

> User        (0b00)
> FIQ         (0b01)
> IRQ         (0b10)
> Supervisor  (0b11)

| | |
|---|---|
| Register 15 | In the 26-bit architectures, the following are also stored in register 15: |

> Four condition flags         (N, Z, C and V)
> The interrupt disable flags   (I and F)
> Two processor mode bits       (M1 and M0)

| | |
|---|---|
| CPSR/SPSR | The 26-bit architectures do not have a CPSR or any SPSRs. |
| Exceptions | An exception (called an address exception) is raised if a memory access instruction uses an address that is greater than $2^{26}$-1 bytes. |

Together, these differences make up the fundamental distinction between 26-bit and 32-bit architectures:

| | |
|---|---|
| 26-bit architectures | all process status (namely the condition flags, interrupt status and processor mode) can be preserved across subroutine calls and nested exceptions without adding any instructions to the entry or exit sequence. |
| 32-bit architectures | give up this functionality to allow 32-bit instruction addresses to be used. |

**ARM Architecture Reference Manual**

ARM  DDI 0100B

## 5.2    Format of Register 15

| 31 30 29 28 27 26 25 | | 2 1 0 |
|---|---|---|
| N Z C V I F | Program Counter | M1 M0 |

Bits[25:2] are collectively known as the Program Counter. Because the Program Counter occupies only 24 bits of register 15, only $2^{24}$ instructions ($2^{26}$ bytes) can be addressed, giving a maximum addressable program size of 64Mbytes.

Bits[31:26] and bits[1:0] are collectively known as the Program Status Register or PSR.

The N, Z, C, V, I, and F bits have the same meaning in both 26-bit and 32-bit architectures.

M[1:0] also have the same meaning in both architectures.

Abort mode and Undef mode are not supported in 26-bit architectures.

Aborts and undefined instruction exceptions have exactly the same actions in both modes, except that in 26-bit architectures, Supervisor mode is entered instead of Abort or Undef mode.

The I, F and M[1:0] bits cannot be written directly when the processor is in User mode; in User mode they are only changed by an exception occurring.

### 5.2.1    Reading register 15

In 26-bit architectures, the value of register 15 is read in five different ways.

1    Most importantly, if register 15 has an unpredictable value in the 32-bit architecture, it also has an unpredictable value when used in the same way in the 26-bit architecture.

2    If register 15 is specified in bits[19:16] of an instruction (and its value is not unpredictable), only the Program Counter (bits[25:2]) are used; all other bits read as zero.

3    If register 15 is specified in bits[3:0] of an instruction (and its value is not unpredictable), all 32 bits are used.

4    If register 15 is stored using STR or STM, the value of the program counter (bits[25:2]) is IMPLEMENTATION DEFINED, but all 32 bits of register are stored.

5    All 32 bits are stored in the Link register (R14) after a Branch with Link instruction or an exception entry.

**ARM Architecture Reference Manual**
ARM  DDI 0100B

## 5.2.2   Writing register 15

In 26-bit architectures the value of register 15, is written in three different ways:

1   Data-processing instructions without the S bit set, Load, and Load Multiple only write the PC part of register 15, and leave the PSR part unchanged.

2   Data-processing instructions with the S bit set and Load Multiple with restore PSR write the PC and the PSR part of register 15.

3   Variants of the CMP, CMN, TST and TEQ instructions write just the PSR part of register 15, and leave the PC part unchanged. These instruction variants are described below.

These read/write rules mean that register 15 is used in three basic ways. It is used as:

- The Rn specifier in data-processing instructions, and as the base address for load and store instructions; only the value of the program counter is used, to simplify PC-relative addressing and position-independent code.

- The Rm specifier in data-processing operands to allow all process status to be restored after a subroutine call or exception by subroutine-return instructions like MOVS PC, LR and LDM..., PC}^. These instructions are unpredictable in user mode on 32-bit architectures, but are legal on 26-bit architectures, as they are used preserve the condition code values across procedure calls.

- The value saved in the Link register to preserve the Program Counter and the PSR across subroutine calls and exceptions.

## 5.3   Writing just the PSR in 26-bit architectures

On 26-bit architectures, the MSR and MRS instructions are not supported. Instead, variants of the CMP, CMN, TST and TEQ instructions are used to write just the PSR part of register 15. These variants are called CMPP, CMNP, TSTP and TEQP, and are distinguished by having instruction bits[15:12] equal to 0b1111 (these bits are usually set to zero for these instructions).

These instructions write their ALU result directly to the PSR part of register 15 (only N, Z, C and V are affected in User mode).

**ARM Architecture Reference Manual**

ARM  DDI 0100B

## 5.4    26-bit PSR Update Instructions

```
TST{<cond>}P   Rn,  <shifter_operand>
TEQ{<cond>}P   Rn,  <shifter_operand>
CMP{<cond>}P   Rn,  <shifter_operand>
CMN{<cond>}P   Rn,  <shifter_operand>
```

**Description**

The instruction is only executed if the condition specified in the instruction matches the condition code status. The conditions are defined in *3.3 The Condition Field* on page 3-4.

The TSTP, TEQP, CMPP and CMNP are 26-bit-only instructions and are used to write the PSR part of register 15 without affecting the PC part of register 15. When the processor is in user mode, only the condition codes will be affected; all other modes allow all PSR bits to be altered.

| 31 | 27 26 25 24 23 22 21 20 19 | 16 15 | 12 11 | 0 |
|----|-----|-----|-----|---|
| cond | 0  0  I  1  0  opc  1 | Rn | SBO | shifter_operand |

**Operation**

```
if ConditionPassed(<cond>) then
        case <opc> of
            0b00 /* TSTP */
                    <alu_out> = Rn AND <shifter_operand>
            0b01 /* TEQP */
                    <alu_out> = Rn EOR <shifter_operand>
            0b10 /* CMP */
                    <alu_out> = Rn - <shifter_operand>
            0b11 /* CMN */
                    <alu_out> = Rn + <shifter_operand>
        endcase
        if R15[1:0] == 0b00 then /* M[1:0] == 0b00, user mode */
            R15[31:28] = <alu_out>[31:28] /* update just NZCV */
        else /* a privileged mode */
            R15[31:26] = <alu_out>[31:26] /* update NZCVIF and ...
*/
            R15[1:0] = <alu_out>[1:0] /* ... update M[1:0] */
```

**Exceptions**

None

**Qualifiers**

Condition Code

**Notes**

**The I bit:** Bit 25 is used to distinguish the immediate and register forms of `<shifter_operand>`. See the data processing instructions for the types of shifter operand.

## 5.5    Address Exceptions

On 26-bit architectures, all data addresses are checked to ensure that they are between 0 and 64 Megabytes (26-bit). If a data address is produced with a 1 in any of the top 6 bits, an address exception is generated. When an address exception is generated, the following actions are performed:

```
R14_svc = address of address exception generating instruction + 4
CPSR[4:0] = 0b00011        ; Supervisor mode
CPSR[7] = 1                ; (Normal) Interrupts disabled
PC = 0x14
```

The address of the instruction which caused the address exception is the value in register 14 minus 8.

### Returning from an address exception

As this exception implies a programming error, it is not usual to return from address exceptions, but if a return is required, use:

```
SUBS PC,R14,#8
```

This restores both the PC and PSR (from R14_svc) and returns to the instruction that generated the address exception.

**ARM Architecture Reference Manual**

ARM DDI 0100B

## 5.6 Backwards Compatibility from 32-bit Architectures

As well as the six (seven in Version 4) 32-bit processor modes, ARM Architecture Version 3 (but not 3G), and 4 (optionally on 4T) implement the four 26-bit processor modes described above, including the register 15 format shown. This allows backwards compatibility for the older 26-bit programs by executing those programs in a 26-bit mode. If the backwards-compatibility support is not implemented, CPSR bit 4 (M[4]) always reads as 1, and all writes are ignored.

The complete list of processor modes is shown in *Table 5-1: 32-bit and 26-bit modes*.

| M[4:0] | Mode | Accessible Registers |
|--------|------|----------------------|
| 0b00000 | User_26 | R0 to R14, PC, (CPSR) |
| 0b00001 | FIQ_26 | R0 to R7, R8_fiq to R14_fiq, PC, (CPSR, SPSR_fiq) |
| 0b00010 | IRQ_26 | R0 to R12, R13_irq, R14_irq, PC, (CPSR, SPSR_irq) |
| 0b00011 | SVC_26 | R0 to R12, R13_svc, R14_svc, PC, (CPSR, SPSR_svc) |
| 0b10000 | User_32 | R0 to R14, PC, CPSR |
| 0b10001 | FIQ_32 | R0 to R7, R8_fiq to R14_fiq, PC, CPSR, SPSR_fiq |
| 0b10010 | IRQ_32 | R0 to R12, R13_irq, R14_irq, PC, CPSR, SPSR_irq |
| 0b10011 | SVC_32 | R0 to R12, R13_svc, R14_svc, PC, CPSR, SPSR_svc |
| 0b10111 | Abort_32 | R0 to R12, R13_abt, R14_abt, PC, CPSR, SPSR_abt |
| 0b11011 | Undef_32 | R0 to R12, R13_und, R14_und, PC, CPSR, SPSR_und |
| 0b11111 | System_32 | R0 to R14, PC, CPSR, (Architecture Version 4 only) |

*Table 5-1: 32-bit and 26-bit modes*

# The 26-bit Architectures

## 5.6.1 32-bit and 26-bit configuration

ARM Architecture Version 3, 3M and 4 (but not 3G or 4T) optionally incorporate two signals that control 32-bit instruction accesses and 32 bit data accesses. The signals are mapped to two bits in register 1 of the system control coprocessor. These signals are:

- PROG32
- DATA32

### 32-bit configuration

1   If PROG32 is active, the processor switches to a 32-bit mode when processing exceptions (including Reset), using the _32 modes for handling all exceptions. This is called a 32-bit configuration. Abort_32 mode is used for handling memory aborts, and Undef_32 for handling undefined instruction exceptions. A 26-bit mode can be selected by putting a 26-bit mode number into the M[4:0] bits of the CPSR (either using MSR or an exception return sequence). A 32-bit mode can also be entered from a 26-bit mode using the MSR instruction. Once in a 26-bit mode, another 26-bit mode can be entered using one of the TEQP, TSTP, CMPP and CMNP instructions, or the MSR instruction.

If an exception occurs when the processor is in a 26-bit mode, only the PC bits from R15[25:2] are copied to the link register; the remaining bits in the link register are zeroed. The PSR bits from R15[31:26] and R15[1:0] are copied into the SPSR, ready for a normal 32-bit return sequence.

2   If PROG32 is active, and DATA32 is not active (32-bit programs with 26-bit data), the result is unpredictable.

### 26-bit configuration

1   If PROG32 is not active, the processor is locked into 26-bit modes (cannot be placed into a 32-bit mode by any means) and handles exceptions in 26-bit modes. This is called a 26-bit configuration. In this configuration, TEQP, TSTP, CMPP and CMNP instructions, or the MSR instruction can be used to switch to 26-bit mode. Attempts to write CPSR bits[4:2] (M[4:2]) are ignored, stopping any attempts to switch to a 32-bit mode, and SVC_26 mode is used to handle memory aborts and undefined instruction exceptions. The program counter is limited to 24 bits, limiting the addressable program memory to 64 Megabytes.

2   If PROG32 is not active, DATA32 has the following actions.

   a)   If DATA32 is not active , all data addresses are checked to ensure that they are between 0 and 64 Megabytes (26 bit). If a data address is produced with a 1 in any of the top 6 bits, an address exception is generated.

   b)   If DATA32 is active, full 32-bit addresses may be produced and are not checked for address exceptions. This allows 26-bit programs to access 32-bit data.

**ARM Architecture Reference Manual**

ARM DDI 0100B

### 5.6.2    Vector exceptions

When the processor is in a 32-bit configuration (PROG32 is active) and in a 26-bit mode (CPSR[4] == 0), data access (but not instruction fetches) to the hard vectors (address 0x0 to 0x1f) cause a data abort, known as a vector exception.

Vector exceptions are always produced if the hard vectors are written in a 32-bit configuration and a 26-bit mode, and it is IMPLEMENTATION DEFINED whether reading the hard vectors in a 32-bit configuration and a 26-bit mode also causes a vector exception.

Vector exceptions are provided to support 26-bit backwards compatibility.
When a vector exception is generated, it indicates that a 26-bit mode process is trying to install a (26-bit) vector handler. Because the processor is in a 32-bit configuration, exceptions will be handled in a 32-bit mode, so a veneer must be used to change from the 32-bit exception mode to a 26-bit mode before calling the 26-bit exception handler. This veneer may be installed on each vector, and can switch to a 26-bit mode before calling any 26-bit handlers.

The 26-bit exception handler's return may also need to be veneered. Some SWI handlers return status information in the processor flags, and this information will need to be transferred from the link register to the SPSR with a return veneer for the SWI handler.

**ARM Architecture Reference Manual**
ARM  DDI 0100B

**6**

# The Thumb Instruction Set

# 6 The Thumb Instruction Set

The Thumb instruction set is a subset of the ARM instruction set.

Thumb is designed to increase the performance of ARM implementations that use a memory data bus, and to allow better code density than ARM. The ARMv4T architecture incorporates both a full 32-bit ARM instruction set and the 16-bit Thumb instruction set. Every Thumb instruction can be encoded in 16 bits.

This chapter lists every Thumb instruction, and gives information on its format and encoding.

# The Thumb Instruction Set

## 6.1 Using this Chapter

This chapter is divided into three parts:

1   Introduction to Thumb
2   Overview of the Thumb instruction types
3   Alphabetical list of instructions

### 6.1.1 Introduction to Thumb *(page 6-3 through 6-4)*

This part describes the Thumb concepts and how it fits in with ARM instruction execution.

### 6.1.2 Overview of the Thumb instruction types *(page 6-5 through 6-15)*

This part describes the functional groups within the Thumb instruction set, and shows relevant examples and encodings. Each functional group lists all its instructions, which you can then find in the alphabetical section. The functional groups are:

1   Branch Instructions
2   Data-processing Instructions
3   Load and Store Register Instructions
4   Load and Store Multiple Instructions

### 6.1.3 Alphabetical list of instructions *(page 6-19 through 6-82)*

This part lists every Thumb instruction in alphabetical order, and gives:

* instruction syntax and functional group
* encoding and operation
* relevant exceptions and qualifiers
* notes on usage

Where relevant, the instruction descriptions show the equivalent ARM instruction and encoding.

**ARM Architecture Reference Manual**

ARM DUI 0100B

## 6.2 Introduction to Thumb

Thumb does not alter the underlying structure of the ARM architecture; it merely presents restricted access to the ARM architecture. All Thumb data-processing instructions operate on full 32-bit values, and full 32-bit bit addresses are produced by both data-access instructions and for instruction fetches.

When the processor is executing Thumb, eight general-purpose integer registers are available, R0 to R7, which are the same physical registers as R0 to R7 when executing ARM. Some Thumb instructions also access the Program Counter (ARM Register 15), the Link Register (ARM Register 14) and the Stack Pointer (ARM Register 13). Further instructions allow limited access to ARM registers 8 to 15 (known as the high registers).

When R15 is read, bit[0] is zero and bits [31:1] contain the PC. When R15 is written, bit[0] is IGNORED and bits[31:1] are written to the PC. Depending on how it is used, the value of the PC is either the address of the instruction plus 4 or is UNPREDICTABLE.

Thumb does not provide direct access to the CPSR or any SPSR (as in the ARM MSR and MRS instructions). Thumb execution is flagged by the T bit (bit 5) in the CPSR:

| | |
|---|---|
| T == 0 | 32-bit instructions are fetched (and the PC is incremented by 4) and are executed as ARM instructions |
| T == 1 | 16-bit instructions are fetched from memory (and the PC is incremented by two) and are executed as Thumb instructions |

### 6.2.1 Entering Thumb state

Thumb execution is normally entered by executing an ARM BX instruction (Branch and eXchange instruction set). This instruction branches to the address held in a general-purpose register, and if bit[0] of that register is 1, Thumb execution begins at the branch target address. If bit[0] of the target register is 0, ARM execution continues from the branch target address.

Thumb execution can also be initiated by setting the T bit in the SPSR and executing an ARM instruction, which restores the CPSR from the SPSR (a data-processing instruction with the S bit set and the PC as the destination, or a Load Multiple and Restore CPSR instruction). This allows an operating system to automatically restart a process independently of whether that process is executing Thumb code or ARM code.

The result is UNPREDICTABLE if the T bit is altered directly by writing the CPSR.

### 6.2.2 Exceptions

Exceptions generated during Thumb execution switch to ARM execution before executing the exception handler (whose first instruction is at the hardware vector). The state of the T bit is preserved in the SPSR, and the LR of the exception mode is set so that the same return instruction performs correctly, regardless of whether the exception occurred during ARM or Thumb execution.

*Table 6-1: Exception return instructions* lists the values of the exception mode LR for exceptions generated during Thumb execution.

# The Thumb Instruction Set

| Exception | Exception Link Register Value | Return Instruction |
|---|---|---|
| Reset | Unpredictable value | MOVS PC, R14 |
| Undefined | Address of Undefined instruction + 2 | MOVS PC, R14 |
| SWI | Address of SWI instruction + 2 | MOVS PC, R14 |
| Prefetch Abort | Address of aborted instruction fetch + 4 | SUBS PC, R14, #4 |
| Data Abort | Address of the instruction that generated the abort + 8 | SUBS PC, R14, #8 |
| IRQ | Address of the next instruction to be executed + 4 | SUBS PC, R14, #4 |
| FIQ | Address of the next instruction to be executed + 4 | SUBS PC, R14, #4 |

*Table 6-1: Exception return instructions*

## 6.3 Instruction Set Overview

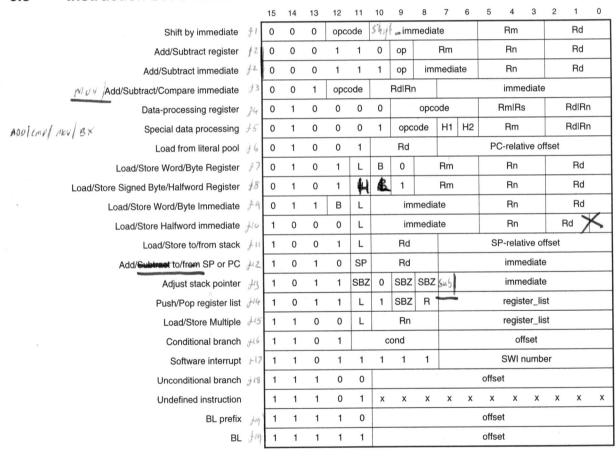

*Figure 6-1: The Thumb instruction set (expanded)*

**ARM Architecture Reference Manual**

ARM DUI 0100B

## 6.4 Branch Instructions

Thumb supports four types of branch instruction.

- an unconditional branch that allows a forward or backward branch of up to 2 Kbytes
- a conditional branch to allow forward and backward branches of up to 256 bytes
- a branch and link (subroutine call) is supported with a pair of instructions that allow forward and backwards branches of up to 4 Mbytes
- a branch and exchange instruction branches to an address in a register and optionally switches to ARM code execution.

### 6.4.1 Encoding

The encoding for these formats is given below:

**Format 1**

```
B<cond>   <target_address>
```

| 15 | 14 | 13 | 12 | 11 | 8 | 7 | 0 |
|----|----|----|----|----|---|---|---|
| 1 | 1 | 0 | 1 | cond | | 8_bit_signed_offset | |

**Format 2**

```
B   <target_address>
```

| 15 | 14 | 13 | 12 | 11 | 10 | 0 |
|----|----|----|----|----|----|---|
| 1 | 1 | 1 | 0 | 0 | 11_bit_signed_offset | |

**Format 3**

```
BL   <target_address>
```

| 15 | 14 | 13 | 12 | 11 | 10 | 0 |
|----|----|----|----|----|----|---|
| 1 | 1 | 1 | 1 | H | 11_bit_signed_offset | |

**Format 4**

```
BX   Rm
```

| 15 | 14 | 13 | 12 | 11 | 10 | 9 | 8 | 7 | 6 | 5 | 3 | 2 | 0 |
|----|----|----|----|----|----|---|---|---|----|---|---|---|---|
| 0 | 1 | 0 | 0 | 0 | 1 | 1 | 0 | 0 | H2 | Rm | | SBZ | |

# The Thumb Instruction Set

## 6.4.2 Examples

```
        B       label       ; unconditionally branch to label
        BCC     label       ; branch to label if carry flags is clear
        BEQ     label       ; branch to label if zero flag is set

        BL      func        ; subroutine call to function

func    .
        .
        MOV     PC, LR      ; R15=R14, return to instruction after the BL

        BX      R12         ; branch to address in R12; begin Thumb execution if
                            ; bit 0 of R12 is zero; otherwise continue executing
                            ; Thumb code
```

## 6.4.3 List of branch instructions

The following instructions follow the formats shown above.

**ARM Architecture Reference Manual**

ARM DUI 0100B

## 6.5    Data-processing Instructions

Thumb data-processing instructions are a subset of the ARM data-processing instructions, as shown below.

All Thumb data-processing instructions set the condition codes.

| Mnemonic | Operation | Action |
|---|---|---|
| MOV Rd, #0 to 255 | Move | Rd := 8-bit immediate |
| MVN Rd, Rm | Move Not | Rd := NOT Rm |
| ADD Rd, Rn, Rm | Add | Rd := Rn + Rm |
| ADD Rd, Rn, #0 to 7 | Add | Rd := Rn + 3-bit immediate |
| ADD Rd, #0 to 255 | Add | Rd := Rd + 8-bit immediate |
| ADC Rd, Rm | Add with Carry | Rd := Rd + Rm + Carry Flag |
| SUB Rd, Rn, Rm | Subtract | Rd := Rn - Rm |
| SUB Rd, Rn, #0 to 7 | Subtract | Rd := Rn - 3-bit immediate |
| SUB Rd, #0 to 255 | Subtract | Rd := Rd - 8-bit immediate |
| SBC Rd, Rm | Subtract with Carry | Rd := Rd - Rm - NOT(Carry Flag) |
| NEG Rd, Rm | Negate | Rd := 0 - Rm |
| AND Rd, Rm | Logical AND | Rd := Rd AND Rm |
| EOR Rd, Rm | Logical Exclusive OR | Rd := Rd EOR Rm |
| ORR Rd, Rm | Logical (inclusive) OR | Rd := Rd OR Rm |
| BIC Rd, Rm | Bit Clear | Rd := Rd AND NOT Rm |
| CMP Rn, #0 to 255 | Compare | update flags after Rn - 8-bit immediate |
| CMP Rn, Rm | Compare | update flags after Rn - Rm |
| CMN Rn, Rm | Compare Negated | update flags after Rn + Rm |
| TST Rn, Rm | Test | update flags after Rn AND Rm |
| MUL Rd, Rs | Multiply | Rd := Rs x Rd |
| LSL Rd, Rm, #0 to 31 | Logical Shift Left | Rd := Rm LSL 5-bit immediate |
| LSL Rd, Rs | Logical Shift Left | Rd := Rd LSL Rs |
| LSR Rd, Rm, #0 to 31 | Logical Shift Right | Rd := Rm LSR 5-bit immediate |
| LSR Rd, Rs | Logical Shift Right | Rd := Rd LSR Rs |
| ASR Rd, Rm, #0 to 31 | Arithmetic Shift Right | Rd := Rm ASR 5-bit immediate |
| ASR Rd, Rs | Arithmetic Shift Right | Rd := Rd ASR Rs |
| ROR Rd, Rs | Rotate Right | Rd := Rd ROR Rs |

*Table 6-2: Thumb data-processing instructions*

# The Thumb Instruction Set

### Examples

```
ADD     R0, R4, R7      ; R0 = R4 + R7
SUB     R6, R1, R2      ; R6 = R1 - R2
ADD     R0, #255        ; R0 = R0 + 255
ADD     R1, R4, #4      ; R1 = R4 + 4
NEG     R3, R1          ; R3 = 0 - R1
AND     R2, R5          ; R2 = R2 AND R5
EOR     R1, R6          ; R1 = R1 EOR R6
CMP     R2, R3          ; update flags after R2 - R3
CMP     R7, #100        ; update flags after R7 - 100
MOV     R0, #200        ; R0 = 200
```

## 6.5.1    High registers

There are seven types of data-processing instruction which operate on ARM registers 8 to 14 and the PC (called the high registers).

| Mnemonic | Operation | Action |
|----------|-----------|--------|
| MOV Rd, Rn | Move | Rd := Rn |
| ADD Rd, Rn | Add | Rd := Rd + Rm |
| CMP Rn, Rm | Compare | update flags after Rn - Rm |
| ADD SP, SP, #0 to 511 | Increment stack pointer | R13 = R13 + #9-bit immediate |
| SUB SP, SP, #0 to 511 | Decrement stack pointer | R13 = R13 - #9-bit immediate |
| ADD Rd, SP, #0 to 1020 | Form Stack address | Rd = R13 + #10-bit immediate |
| ADD Rd, PC, #0 to 1020 | Form PC address | Rd = PC + #10-bit immediate |

*Table 6-3: High register data-processing instructions*

### Examples

```
MOV     R0, R12         ; R0 = R12
ADD     R10, R1, R2     ; R6 = R1 - R2
MOV     PC, LR          ; PC = R14
CMP     R10, R11        ; update flags after R10 - R11
SUB     SP, SP, #10     ; increase stack size by 100 bytes
ADD     SP, SP, #16     ; decrease stack size by 16 bytes
ADD     R2, SP, #20     ; R2 = SP + 20
ADD     R0, PC, #500    ; R0 = PC + 500
```

**ARM Architecture Reference Manual**

ARM DUI 0100B

## 6.5.2    Formats

Data-processing instructions perform an operation on the general processor registers:

### Format 1

```
<opcode1>  Rd, Rn, Rm
<opcode1> := ADD | SUB
```

| 15 | 14 | 13 | 12 | 11 | 10 | 9 | 8 | | 6 | 5 | | 3 | 2 | | 0 |
|----|----|----|----|----|----|---|---|---|---|---|---|---|---|---|---|
| 0 | 0 | 0 | 1 | 1 | 0 | op_1 | Rm | | | Rn | | | Rd | | |

### Format 2

```
<opcode2>  Rd, Rn, <3_bit_immed>
<opcode2> := ADD | SUB
```

| 15 | 14 | 13 | 12 | 11 | 10 | 9 | 8 | | 6 | 5 | | 3 | 2 | | 0 |
|----|----|----|----|----|----|---|---|---|---|---|---|---|---|---|---|
| 0 | 0 | 0 | 1 | 1 | 1 | op_2 | 3_bit_immediate | | | Rn | | | Rd | | |

### Format 3

```
<opcode3>  Rd|Rn, #<8_bit_immed>
<opcode3> := ADD | SUB | MOV | CMP
```

| 15 | 14 | 13 | 12 | 11 | 10 | | 8 | 7 | | | | | | | 0 |
|----|----|----|----|----|----|---|---|---|---|---|---|---|---|---|---|
| 0 | 0 | 1 | op_3 | | Rd|Rn | | | 8_bit_immediate | | | | | | | |

### Format 4

```
<opcode4>  Rd, Rn #<shift_imm>
<opcode4> := LSL | LSR | ASR
```

| 15 | 14 | 13 | 12 | 11 | 10 | | | | 6 | 5 | | 3 | 2 | | 0 |
|----|----|----|----|----|----|---|---|---|---|---|---|---|---|---|---|
| 0 | 0 | 0 | op_4 | | shift_immediate | | | | | Rm | | | Rd | | |

### Format 5

```
<opcode5>  Rd | Rn, Rm | Rs
<opcode5> := MVN|CMP|CMN|TST|ADC|SBC|NEG|MUL|
             LSL|LSR|ASR|ROR|AND|EOR|ORR|BIC
```

| 15 | 14 | 13 | 12 | 11 | 10 | 9 | | | 6 | 5 | | 3 | 2 | | 0 |
|----|----|----|----|----|----|---|---|---|---|---|---|---|---|---|---|
| 0 | 1 | 0 | 0 | 0 | 0 | | op_5 | | | Rm|Rs | | | Rd|Rn | | |

# The Thumb Instruction Set

**Format 6**

```
ADD  Rd, <reg>, #<8_bit_immed>
<reg> := SP | PC
```

| 15 | 14 | 13 | 12 | 11 | 10 | 8 | 7 | 0 |
|----|----|----|----|----|----|---|---|---|
| 1 | 0 | 1 | 0 | reg | Rd | | 8_bit_immediate | |

**Format 7**

```
<opcode6>  SP, SP, #<7_bit_immed>
<opcode6> := ADD | SUB
```

| 15 | 14 | 13 | 12 | 11 | 10 | 9 | 8 | 7 | 6 | 0 |
|----|----|----|----|----|----|---|---|---|---|---|
| 1 | 0 | 1 | 1 | 0 | 0 | 0 | 0 | op_6 | 7_bit_immediate | |

## 6.5.3 List of data-processing instructions

The following instructions follow the formats shown above.

**ARM Architecture Reference Manual**

ARM DUI 0100B

# The Thumb Instruction Set

## 6.6    Load and Store Register Instructions

Thumb supports 8 types of load and store register instructions. Two basic addressing modes are available:

- register plus register
- register plus 5-bit immediate

If an immediate offset is used, it is scaled by 4 for word access and 2 for halfword accesses. Three special instructions allow a load using the PC as a base with a 1 Kbyte (word-aligned) immediate offset, and a load and store instructions with the stack pointer (R13) as the base and a 1Kbyte (word aligned) immediate offset.

### 6.6.1    Formats

Load and Store instructions perform an operation on the general processor registers. Load and store instructions have the following formats:

**Format 1**
```
<opcode1>  Rd, [Rn, #<5_bit_offset>]
<opcode1> := LDR|LDRH|LDRB|STR|STRH|STRB
```

| 15 | 11 | 10 | 6 | 5 | 3 | 2 | 0 |
|----|----|----|---|---|---|---|---|
| opcode1 | | 5_bit_offset | | Rn | | Rd | |

**Format 2**
```
<opcode2>  Rd, [Rn, Rm]
<opcode2> := LDR|LDRH|LDRSH|LDRB|LDRSB|STR|STRH|STRB
```

| 15 | 9 | 8 | 6 | 5 | 3 | 2 | 0 |
|----|---|---|---|---|---|---|---|
| opcode2 | | Rm | | Rn | | Rd | |

**Format 3**
```
LDR  Rd, [PC, #<8_bit_offset>]
```

| 15 | 14 | 13 | 12 | 11 | 10 | 8 | 7 | 0 |
|----|----|----|----|----|----|---|---|---|
| 0 | 1 | 0 | 0 | 1 | Rd | | 8_bit_immediate | |

**Format 4**
```
<opcode3>  Rd, [SP, #<8_bit_offset>]
<opcode3>  := LDR | STR
```

| 15 | 14 | 13 | 12 | 11 | 10 | 8 | 7 | 0 |
|----|----|----|----|----|----|---|---|---|
| 1 | 0 | 0 | 1 | L | Rd | | 8_bit_immediate | |

**ARM Architecture Reference Manual**
ARM DUI 0100B

## 6.6.2    Examples

```
LDR    R4, [R2, #4]    ; Load word into R4 from address R2 + 4
LDR    R4, [R2, R1]    ; Load word into R4 from address R2 + R1
STR    R0, [R7, #0x7c] ; Store word from R0 to address R7 + 124
STRB   R1, [R5, #31]   ; Store byte from R1 to address R5 + 31
STRH   R4, [R2, R3]    ; Store halfword from R4 to R2 + R3
LDRSB  R5, [R0, #0]    ; Load signed byte into R5 byte from R0
LDRSH  R1, [R2, #10]   ; Load signed halfword to R1 from R2 + 10
LDRH   R3, [R6, R5]    ; Load word into R3 from R6 + R5
LDRB   R2, [R1, #5]    ; Load byte into R2 from R1 + 5
LDR    R6, [PC, #0x3fc]; Load R6 from PC + 0x3fc
LDR    R5, [SP, #64]   ; Load R5 from SP + 64
STR    R4, [SP, #0x260]; Load R5 from SP + 0x260
```

## 6.6.3    List of load and store register instructions

The following instructions follow the formats shown above.

# The Thumb Instruction Set

## 6.7　Load and Store Multiple Instructions

Thumb supports four types of load and store multiple instructions.

Two (a load and a store) are designed to support block copy; they have a fixed increment-after addressing mode from a base register.

The other two instructions (called PUSH and POP) also have a fixed addressing mode. They implement a full descending stack, and the stack pointer (R13) is used as the base register.

All four instructions can transfer any or all of the lower 8 registers. PUSH can also stack the return address and POP can load the PC. All four instructions update the base register after the transfer.

### 6.7.1　Formats

**Format 1**

```
<opcode1>  Rn!, <register_list>
<opcode1>  := LDMIA | STMIA
```

| 15 | 14 | 13 | 12 | 11 | 10 | 8 | 7 | 0 |
|----|----|----|----|----|----|----|----|----|
| 1 | 1 | 0 | 0 | L | Rn | | register_list | |

**Format 2**

```
POP    {<register_list>,<PC>}
PUSH   {<register_list>,<LR>}
```

| 15 | 14 | 13 | 12 | 11 | 10 | 9 | 8 | 7 | 0 |
|----|----|----|----|----|----|----|----|----|----|
| 1 | 0 | 1 | 1 | 1 | 1 | L | R | register_list | |

### 6.7.2　Examples

```
LDMIA   R7!, {R0 - R3, R5}      ; Load R0 to R3 and R5 from R7
                                ; then add 20 to R7
STMIA   R0!, {R3, R4, R5}       ; Store R3-R5 to R0: add 12 to R0 function
STMFD   R13!, {R0-R7, LR}       ; save all regs and return address
  .                             ; code of the function body
  .
STMFD   R13!, {R0-R7, PC}       ; restore all register and return

PUSH    {R0 - R7, LR}           ; push onto the stack (R13) R0 - R7 and
                                ; the return address
POP     {R0 - R7, PC}           ; restore R0 - R7 from the stack
                                ; and the program counter, and return
```

**ARM Architecture Reference Manual**

ARM DUI 0100B

## 6.7.3    List of load and store multiple instructions

The following instructions follow the formats shown above.

| | | |
|---|---|---|
| LDM | Load multiple | page 6-40 |
| POP | Pop multiple | page 6-62 |
| PUSH | Push multiple | page 6-64 |
| STM | Store multiple | page 6-68 |

# Thumb Instructions

**Instruction name**
given in the following alphabetical list

**Description**

**Syntax**

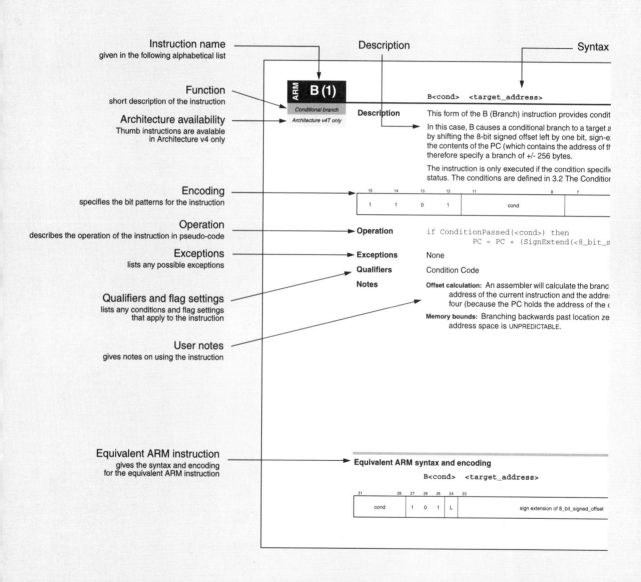

**Function**
short description of the instruction

**Architecture availability**
Thumb instructions are avalable
in Architecture v4 only

**Encoding**
specifies the bit patterns for the instruction

**Operation**
describes the operation of the instruction in pseudo-code

**Exceptions**
lists any possible exceptions

**Qualifiers and flag settings**
lists any conditions and flag settings
that apply to the instruction

**User notes**
gives notes on using the instruction

**Equivalent ARM instruction**
gives the syntax and encoding
for the equivalent ARM instruction

ARM **B (1)**

*Conditional branch*

*Architecture v4T only*

B<cond>   <target_address>

**Description**

This form of the B (Branch) instruction provides condit

In this case, B causes a conditional branch to a target a
by shifting the 8-bit signed offset left by one bit, sign-e:
the contents of the PC (which contains the address of th
therefore specify a branch of +/- 256 bytes.

The instruction is only executed if the condition specifi
status. The conditions are defined in 3.2 The Condition

| 15 | 14 | 13 | 12 | 11 | | 8 | 7 |
|----|----|----|----|----|----|----|----|
| 1 | 1 | 0 | 1 | | cond | | |

**Operation**

```
if ConditionPassed(<cond>) then
        PC = PC + (SignExtend(<8_bit_s
```

**Exceptions**    None

**Qualifiers**    Condition Code

**Notes**

**Offset calculation:** An assembler will calculate the branc
address of the current instruction and the addre:
four (because the PC holds the address of the c

**Memory bounds:** Branching backwards past location ze
address space is UNPREDICTABLE.

**Equivalent ARM syntax and encoding**

B<cond>   <target_address>

| 31 | 28 | 27 | 26 | 25 | 24 | 23 | |
|----|----|----|----|----|----|----|----|
| cond | | 1 | 0 | 1 | L | | sign extension of 8_bit_signed_offset |

| Description | The ADC (Add with Carry) instruction is used to synthesize 64-bit addition. If register pairs R0,R1 and R2,R3 hold 64-bit values (R0 and R2 hold the least-significant word), the following instructions leave the 64-bit sum in R0,R1: |
|---|---|

```
ADD    R0,R2
ADC    R1,R3
```

The instruction ADC R0,R0 produces a single-bit Rotate Left with Extend operation (33-bit rotate though the carry flag) on R0.

ADC adds the value of register Rd, and the value of the Carry flag, and the value of register Rm, and stores the result in register Rd. The condition code flags are updated (based on the result).

| 15 | 14 | 13 | 12 | 11 | 10 | 9 | 8 | 7 | 6 | 5 | | | 3 | 2 | | | 0 |
|----|----|----|----|----|----|----|----|----|----|----|----|----|----|----|----|----|----|
| 0 | 1 | 0 | 0 | 0 | 0 | 0 | 1 | 0 | 1 | | Rm | | | | Rd | | |

**Operation**

```
Rd = Rd + Rm + C Flag
N Flag = Rd[31]
Z Flag = if Rd == 0 then 1 else 0
C Flag = CarryFrom(Rd + Rm + C Flag)
V Flag = OverflowFrom(Rd + Rm + C Flag)
```

**Exceptions**   None

**Qualifiers**   None

### Equivalent ARM syntax and encoding

```
ADCS Rd, Rd, Rm
```

| 31 | 30 | 29 | 28 | 27 | 26 | 25 | 24 | 23 | 22 | 21 | 20 | 19 | | | 16 | 15 | | | 12 | 11 | 10 | 9 | 8 | 7 | 6 | 5 | 4 | 3 | | | 0 |
|----|----|----|----|----|----|----|----|----|----|----|----|----|----|----|----|----|----|----|----|----|----|----|----|----|----|----|----|----|----|----|----|
| 1 | 1 | 1 | 0 | 0 | 0 | 0 | 0 | 1 | 0 | 1 | 1 | | Rd | | | | Rd | | | 0 | 0 | 0 | 0 | 0 | 0 | 0 | 0 | | Rm | | |

# ADD (1)

*Add immediate*

*Architecture v4T only*

**Description**  This form of the ADD instruction adds a small constant value to the value of a register and stores the result in a second register.

In this case, ADD adds the value of register Rn and the value of the 3-bit immediate (values 0 to 7), and stores the result in the destination register Rd. The condition code flags are updated (based on the result).

| 15 | 14 | 13 | 12 | 11 | 10 | 9 | 8      6 | 5      3 | 2      0 |
|----|----|----|----|----|----|----|----|----|----|
| 0 | 0 | 0 | 1 | 1 | 1 | 0 | 3_bit_immediate | Rn | Rd |

**Operation**
```
Rd = Rn + <3_bit_immediate>
N Flag = Rd[31]
Z Flag = if Rd == 0 then 1 else 0
C Flag = CarryFrom(Rn + <3_bit_immed>)
V Flag = OverflowFrom(Rn + <3_bit_immed>)
```

**Exceptions**  None

**Qualifiers**  None

---

**Equivalent ARM syntax and encoding**

        ADDS Rd, Rn, #<3_bit_immediate>

| 31 30 29 28 | 27 26 25 24 23 22 21 20 | 19     16 | 15     12 | 11 10 9 8 7 6 5 4 3 2 | 0 |
|----|----|----|----|----|----|
| 1 1 1 0 | 0 0 1 0 1 0 0 1 | Rn | Rd | 0 0 0 0 0 0 0 0 0 | #3_bit_imm |

**ARM Architecture Reference Manual**
ARM DUI 0100B

`ADD Rd, #<8_bit_immediate>`

**Description**  This form of the ADD instruction is also used to add a large constant value to the value of a register and to store the result back in the same register.

In this case, ADD instruction adds the value of register Rd and the value of the 8-bit immediate (values 0 to 255), and stores the result back in register Rd. The condition code flags are updated (based on the result).

| 15 | 14 | 13 | 12 | 11 | 10 | 9 | 8 | 7 | | | | | | | 0 |
|----|----|----|----|----|----|---|---|---|---|---|---|---|---|---|---|
| 0 | 0 | 1 | 1 | 0 | | Rd | | | | | 8_bit_immediate | | | | |

**Operation**
```
Rd = Rd + <8_bit_immed>
N Flag = Rd[31]
Z Flag = if Rd == 0 then 1 else 0
C Flag = CarryFrom(Rd + <8_bit_immed>)
V Flag = OverflowFrom(Rd + <8_bit_immed>)
```

**Exceptions**  None

**Qualifiers**  None

**Equivalent ARM syntax and encoding**

```
ADDS Rd, Rd, #<8_bit_immediate>
```

| 31 | 30 | 29 | 28 | 27 | 26 | 25 | 24 | 23 | 22 | 21 | 20 | 19 | | | | 16 | 15 | | | | 12 | 11 | 10 | 9 | 8 | 7 | | | | | 0 |
|----|----|----|----|----|----|----|----|----|----|----|----|----|---|---|---|----|----|---|---|---|----|----|----|---|---|---|---|---|---|---|---|
| 1 | 1 | 1 | 0 | 0 | 0 | 1 | 0 | 1 | 0 | 0 | 1 | | Rd | | | | | Rd | | | | 0 | 0 | 0 | 0 | | | 8_bit_immediate | | | |

**ARM Architecture Reference Manual**
ARM DUI 0100B

# ADD (3)

`ADD Rd, Rn, Rm`

**Add register**

**Architecture v4T only**

**Description**    This form of the ADD instruction adds the value of one register to the value of a second register, and stores the result in a third register.

In this case, ADD adds the value of register Rn and the value of register Rm, and stores the result in the destination register Rd. The condition code flags are updated (based on the result).

| 15 | 14 | 13 | 12 | 11 | 10 | 9 | 8 | | 6 | 5 | | 3 | 2 | | 0 |
|----|----|----|----|----|----|---|---|---|---|---|---|---|---|---|---|
| 0 | 0 | 0 | 1 | 1 | 0 | 0 | | Rm | | | Rn | | | Rd | |

**Operation**
```
Rd = Rn + Rm
N Flag = Rd[31]
Z Flag = if Rd == 0 then 1 else 0
C Flag = CarryFrom(Rn + Rm)
V Flag = OverflowFrom(Rn + Rm)
```

**Exceptions**    None

**Qualifiers**    None

**Equivalent ARM syntax and encoding**

`ADDS Rd, Rn, Rm`

| 31 | 30 | 29 | 28 | 27 | 26 | 25 | 24 | 23 | 22 | 21 | 20 | 19 | | 16 | 15 | | 12 | 11 | 10 | 9 | 8 | 7 | 6 | 5 | 4 | 3 | | 0 |
|----|----|----|----|----|----|----|----|----|----|----|----|----|---|----|----|---|----|----|----|---|---|---|---|---|---|---|---|---|
| 1 | 1 | 1 | 0 | 0 | 0 | 0 | 0 | 1 | 0 | 0 | 1 | | Rn | | | Rd | | 0 | 0 | 0 | 0 | 0 | 0 | 0 | 0 | | Rm | |

**ARM Architecture Reference Manual**
ARM DUI 0100B

ADD Rd, Rm

**Description**  This form of the ADD instruction is used for addition of values in the high registers. In this case, ADD:

- adds the value of a low register to a high register (H1=1, H2=0), or
- adds the value of a high register to a low register (H1=0, H2=1), or
- adds the value of a high register to another high register (H1=1, H2=1)

The condition code flags are not affected.

| 15 | 14 | 13 | 12 | 11 | 10 | 9 | 8 | 7 | 6 | 5 | | 3 | 2 | | 0 |
|----|----|----|----|----|----|---|---|---|---|---|---|---|---|---|---|
| 0 | 1 | 0 | 0 | 0 | 1 | 0 | 0 | H1 | H2 | | Rm | | | Rd | |

**Operation**  Rd = Rd + Rm

**Exceptions**  None

**Qualifiers**  None

**Notes**  **Operand restriction:** If a low register is specified for Rd and Rm (H1=0 and H2=0), the result is UNPREDICTABLE.

**Equivalent ARM syntax and encoding**

ADD Rd, Rd, Rm

| 31 | 30 | 29 | 28 | 27 | 26 | 25 | 24 | 23 | 22 | 21 | 20 | 19 | 18 | | 16 | 15 | 14 | | 12 | 11 | 10 | 9 | 8 | 7 | 6 | 5 | 4 | 3 | 2 | | 0 |
|----|----|----|----|----|----|----|----|----|----|----|----|----|----|---|----|----|----|---|----|----|----|---|---|---|---|---|---|---|---|---|---|
| 1 | 1 | 1 | 0 | 0 | 0 | 0 | 0 | 1 | 0 | 0 | 0 | H1 | | Rd | | H1 | | Rd | | 0 | 0 | 0 | 0 | 0 | 0 | 0 | 0 | H2 | | Rm | |

# ADD (5)

`ADD Rd, PC, #<8_bit_immediate>)`

**Add immediate
to program counter**

**Architecture v4T only**

**Description**    This form of the ADD instruction is used to address a PC-relative (word-sized) variable.

In this case, ADD clears the bottom two bits of the value of the PC and adds the result to the value of the 8-bit immediate (values 0 to 255), and stores the result in register Rd.    *Shifted left 2 places*

| 15 | 14 | 13 | 12 | 11 | 10      8 | 7                           0 |
|----|----|----|----|----|-----------|-------------------------------|
| 1  | 0  | 1  | 0  | 0  | Rd        | 8_bit_immediate               |

**Operation**    Rd = (PC AND 0xfffffffc) + (8_bit_immed> <<2)

**Exceptions**    None

**Qualifiers**    None

**Equivalent ARM syntax and encoding**

`ADD Rd, PC, #<8_bit_immediate>`

*<LSL 2*

| 31 | 30 | 29 | 28 | 27 | 26 | 25 | 24 | 23 | 22 | 21 | 20 | 19 | 18 | 17 | 16 | 15          12 | 11 | 10 | 9 | 8 | 7                     0 |
|----|----|----|----|----|----|----|----|----|----|----|----|----|----|----|----|----------------|----|----|---|---|-------------------------|
| 1  | 1  | 1  | 0  | 0  | 0  | 1  | 0  | 1  | 0  | 0  | 0  | 1  | 1  | 1  | 1  | Rd             | 0  | 0  | 0 | 0 | 8_bit_immediate         |

**ARM Architecture Reference Manual**

ARM DUI 0100B

ARM

| **Description** | This form of the ADD instruction is used to address an SP-relative (word-sized) variable. |
|---|---|
| | In this case, ADD adds the value of the SP and the value of the 8-bit immediate (values 0 to 255), and stores the result in register Rd. |

LSL #2

| 15 | 14 | 13 | 12 | 11 | 10 | 8 | 7 | | 0 |
|---|---|---|---|---|---|---|---|---|---|
| 1 | 0 | 1 | 0 | 1 | Rd | | | 8_bit_immediate | |

| **Operation** | Rd = SP + (<8_bit_immed> << 2) |
|---|---|
| **Exceptions** | None |
| **Qualifiers** | None |

**Equivalent ARM syntax and encoding**

ADD Rd, SP, #<8_bit_immediate>

LSL 2

| 31 | 30 | 29 | 28 | 27 | 26 | 25 | 24 | 23 | 22 | 21 | 20 | 19 | 18 | 17 | 16 | 15 | | 12 | 11 | 10 | 9 | 8 | 7 | | 0 |
|---|---|---|---|---|---|---|---|---|---|---|---|---|---|---|---|---|---|---|---|---|---|---|---|---|---|
| 1 | 1 | 1 | 0 | 0 | 0 | 1 | 0 | 1 | 0 | 0 | 0 | 1 | 1 | 0 | 1 | | Rd | | 0 | 0 | 0 | 0 | | 8_bit_immediate | |

*Increment
stack pointer*

*Architecture v4T only*

ADD SP, SP, #<7_bit_immediate>

**Description**

This form of the ADD instruction is used to decrease the size of the stack.

In this case, ADD adds the value of the SP and the value of the 7-bit immediate (values 0 to 127), and stores the result back in the SP.

*O to IFC*

| 15 | 14 | 13 | 12 | 11 | 10 | 9 | 8 | 7 | 6 | | | 0 |
|----|----|----|----|----|----|---|---|---|---|---|---|---|
| 1 | 0 | 1 | 1 | 0 | 0 | 0 | 0 | 0 | | 7_bit_immediate | | |

**Operation**

SP = SP + <7_bit_immed>  *<< 2*

**Exceptions**

None

**Qualifiers**

None

**Equivalent ARM syntax and encoding**

ADD SP, SP, #<7_bit_immediate>

| 31 | 30 | 29 | 28 | 27 | 26 | 25 | 24 | 23 | 22 | 21 | 20 | 19 | 18 | 17 | 16 | 15 | 14 | 13 | 12 | 11 | 10 | 9 | 8 | 7 | 6 | | 0 |
|----|----|----|----|----|----|----|----|----|----|----|----|----|----|----|----|----|----|----|----|----|----|---|---|---|---|---|---|
| 1 | 1 | 1 | 0 | 0 | 0 | 1 | 0 | 1 | 0 | 0 | 0 | 1 | 1 | 0 | 1 | 1 | 1 | 0 | 1 | 0 | 0 | 0 | 0 | 0 | | 7_bit_immediate | |

**ARM Architecture Reference Manual**
ARM DUI 0100B

`AND Rd, Rm`

**Description**  The AND (Logical AND) instruction is most useful for extracting a field from a register, by ANDing the register with a mask value that has 1's in the field to be extracted, and 0's elsewhere.

AND performs a bitwise AND of the value of register Rm with the value of register Rd, and stores the result back in register Rd. The condition code flags are updated (based on the result).

| 15 | 14 | 13 | 12 | 11 | 10 | 9 | 8 | 7 | 6 | 5 | | 3 | 2 | | 0 |
|----|----|----|----|----|----|---|---|---|---|---|---|---|---|---|---|
| 0 | 1 | 0 | 0 | 0 | 0 | 0 | 0 | 0 | 0 | | Rm | | | Rd | |

**Operation**
```
Rd = Rd AND Rm
N Flag = Rd[31]
Z Flag = if Rd == 0 then 1 else 0
C Flag = <shifter_carry_out>
V Flag = unaffected
```

**Exceptions**  None

**Qualifiers**  None

---

**Equivalent ARM syntax and encoding**

`ANDS Rd, Rd, Rm`

| 31 | 30 | 29 | 28 | 27 | 26 | 25 | 24 | 23 | 22 | 21 | 20 | 19 | | 16 | 15 | | 12 | 11 | 10 | 9 | 8 | 7 | 6 | 5 | 4 | 3 | | 0 |
|----|----|----|----|----|----|----|----|----|----|----|----|----|---|----|----|---|----|----|----|---|---|---|---|---|---|---|---|---|
| 1 | 1 | 1 | 0 | 0 | 0 | 0 | 0 | 0 | 0 | 0 | 1 | | Rd | | | Rd | | 0 | 0 | 0 | 0 | 0 | 0 | 0 | 0 | | Rm | |

**ARM Architecture Reference Manual**
ARM DUI 0100B

## ASR (1)

*Arithmetic shift right (immediate)*

*Architecture v4T only*

`ASR Rd, Rm, #<shift_imm>`

The ASR (Arithmetic Shift Right) instruction is used to provide the signed value of a register divided by a constant power of 2.

In this case, ASR performs an arithmetic shift right of the value of register Rm by an immediate value in the range 1 to 32, and stores the result into the destination register Rd. The sign bit of Rm (Rm[31]) is inserted into the vacated bit positions.

A shift by 32 is encoded by:

    <shift_imm> = 0

The condition code flags are updated (based on the result).

| 15 14 13 | 12 | 11 | 10 ... 6 | 5 ... 3 | 2 ... 0 |
|---|---|---|---|---|---|
| 0 0 0 | 1 | 0 | shift_imm | Rm | Rd |

**Operation**

```
if <shift_imm> == 0
        if Rm[31] == 0 then
            C Flag = Rm[31]
            Rd = 0
        else /* Rd[31] == 1 */
            C Flag = Rm[31]
            Rd = 0xffffffff
else /* <shift_imm> > 0 */
        C Flag = Rm[<shift_imm> - 1]
        Rd = Rm Arithmetic_Shift_Right <shift_imm>
N Flag = Rd[31]
Z Flag = if Rd == 0 then 1 else 0
V Flag = unaffected
```

**Exceptions**   None

**Qualifiers**   None

**Equivalent ARM syntax and encoding**

    MOVS Rd, Rm, ASR #<shift_imm>

| 31 30 29 28 | 27 26 25 24 23 22 21 20 | 19 ... 16 | 15 ... 12 | 11 ... 7 | 6 5 4 | 3 ... 0 |
|---|---|---|---|---|---|---|
| 1 1 1 0 | 0 0 0 1 1 0 1 1 | SBZ | Rd | shift_imm | 1 0 0 | Rm |

**ARM Architecture Reference Manual**
ARM DUI 0100B

| | | | | | |
|---|---|---|---|---|---|

**Description**

This form of the ASR (Arithmetic Shift Right) instruction is used to provide the signed value of a register divided by a constant power of 2.

In this case, ASR performs an arithmetic shift right of the value of register Rd by the value in the least-significant byte of register Rs, and stores the result back into the register Rd. The sign bit of the original Rd (Rd[31]) is inserted into the vacated bit positions. The condition code flags are updated (based on the result).

| 15 | 14 | 13 | 12 | 11 | 10 | 9 | 8 | 7 | 6 | 5 | 3 | 2 | 0 |
|----|----|----|----|----|----|---|---|---|---|---|---|---|---|
| 0 | 1 | 0 | 0 | 0 | 0 | 0 | 1 | 0 | 0 | Rs | | Rd | |

**Operation**

```
if Rs[7:0] == 0 then
        C Flag = unaffected
        Rd = unaffected
else if Rs[7:0] < 32 then
        C Flag = Rd[Rs[7:0] - 1]
        Rd = Rd Arithmetic_Shift_Right Rs[7:0]
else /* Rs[7:0] >= 32 */
        if Rd[31] == 0 then
            C Flag = Rd[31]
            Rd = 0
        else /* Rd[31] == 1 */
            C Flag = Rd[31]
            Rd = 0xffffffff
```

**Exceptions**    None

**Qualifiers**    None

**Equivalent ARM syntax and encoding**

MOVS Rd, Rd, ASR Rs

| 31 | 30 | 29 | 28 | 27 | 26 | 25 | 24 | 23 | 22 | 21 | 20 | 19 | 16 | 15 | 12 | 11 | 8 | 7 | 6 | 5 | 4 | 3 | 0 |
|----|----|----|----|----|----|----|----|----|----|----|----|----|----|----|----|----|---|---|---|---|---|---|---|
| 1 | 1 | 1 | 0 | 0 | 0 | 0 | 1 | 1 | 0 | 1 | 1 | SBZ | | Rd | | Rs | | 0 | 1 | 0 | 1 | Rd | |

**ARM Architecture Reference Manual**
ARM DUI 0100B

# B (1)

B<cond>   <target_address>

*Conditional branch*

*Architecture v4T only*

**Description**   This form of the B (Branch) instruction provides conditional changes to program flow.

In this case, B causes a conditional branch to a target address. The branch target address is calculated by shifting the 8-bit signed offset left by one bit, sign-extending the result to 32 bits, and adding this to the contents of the PC (which contains the address of the branch instruction plus 4). The instruction can therefore specify a branch of +/- 256 bytes.

The instruction is only executed if the condition specified in the instruction matches the condition code status. The conditions are defined in *3.3 The Condition Field* on page 3-4.

| 15 | 14 | 13 | 12 | 11          8 | 7                                    0 |
|----|----|----|----|---------------|----------------------------------------|
| 1  | 1  | 0  | 1  | cond          | 8_bit_signed_offset                    |

**Operation**   if ConditionPassed(<cond>) then
                    PC = PC + (SignExtend(<8_bit_signed_offset>) << 1)

**Exceptions**   None

**Qualifiers**   Condition Code

**Notes**   **Offset calculation:** An assembler will calculate the branch offset address from the difference between the address of the current instruction and the address of the target (given as a program label) minus four (because the PC holds the address of the current instruction plus four).

**Memory bounds:** Branching backwards past location zero and forwards over the end of the 32-bit address space is UNPREDICTABLE.

**Equivalent ARM syntax and encoding**

B<cond>   <target_address>

| 31        28 | 27 26 25 24 | 23                                  8 | 7                          0 |
|--------------|-------------|---------------------------------------|------------------------------|
| cond         | 1  0  1  0  | sign extension of 8_bit_signed_offset | 8_bit_signed_offset          |

**ARM Architecture Reference Manual**
ARM DUI 0100B

B   <target_address>

**Description**    This form of the B (Branch) instruction provides unconditional changes to program flow.

In this case, B causes an unconditional branch to a target address. The branch target address is calculated by shifting the 11-bit signed (two's complement) offset left one bit, sign-extending the result to 32 bits, and adding this to the contents of the PC (which contains the address of the branch instruction plus 4).
The instruction can therefore specify a branch of +/- 2048 bytes.

| 15 | 14 | 13 | 12 | 11 | 10                                     0 |
|----|----|----|----|----|----|
| 1 | 1 | 1 | 0 | 0 | 11_bit_signed_offset |

**Operation**    PC = PC + (SignExtend(<11_bit_signed_offset>) << 1)

**Exceptions**    None

**Qualifiers**    None

**Notes**    **Offset calculation:** An assembler will calculate the branch offset address from the difference between the address of the current instruction and the address of the target (given as a program label) minus four (because the PC holds the address of the current instruction plus four).

**Memory bounds:** Branching backwards past location zero and forwards over the end of the 32-bit address space is UNPREDICTABLE.

**Equivalent ARM syntax and encoding**

B   <target_address>

| 31              28 | 27 | 26 | 25 | 24 | 23                         12 | 11 | 10                       0 |
|----|----|----|----|----|----|----|----|
| 1 1 1 0 <br> cond | 1 | 0 | 1 | 0 | sign extension of 11_bit_signed_offset | | 11_bit_signed_offset |

# BIC

*Bit clear*

*Architecture v4T only*

BIC Rd, Rm

**Description**

The BIC (Bit Clear) instruction can be used to clear selected bits in a register. For each bit, BIC with 1 clears the bit, and BIC with 0 leaves it unchanged.

BIC performs a bitwise AND of the complement of the value of register Rm with the value of register Rd, and stores the result back in register Rd. The condition code flags are updated (based on the result).

| 15 | 14 | 13 | 12 | 11 | 10 | 9 | 8 | 7 | 6 | 5 | | 3 | 2 | | 0 |
|----|----|----|----|----|----|---|---|---|---|---|---|---|---|---|---|
| 0 | 1 | 0 | 0 | 0 | 0 | 1 | 1 | 1 | 0 | | Rm | | | Rd | |

**Operation**

```
Rd = Rd AND NOT Rm
N Flag = Rd[31]
Z Flag = if Rd == 0 then 1 else 0
C Flag = <shifter_carry_out>
V Flag = unaffected
```

**Exceptions**   None

**Qualifiers**   None

**Equivalent ARM syntax and encoding**

BICS Rd, Rd, Rm

| 31 | 30 | 29 | 28 | 27 | 26 | 25 | 24 | 23 | 22 | 21 | 20 | 19 | | 16 | 15 | | 12 | 11 | 10 | 9 | 8 | 7 | 6 | 5 | 4 | 3 | | | 0 |
|----|----|----|----|----|----|----|----|----|----|----|----|----|---|----|----|---|----|----|----|---|---|---|---|---|---|---|---|---|---|
| 1 | 1 | 1 | 0 | 0 | 0 | 0 | 1 | 1 | 1 | 0 | 1 | | Rd | | | Rd | | 0 | 0 | 0 | 0 | 0 | 0 | 0 | 0 | | Rm | | |

**ARM Architecture Reference Manual**

ARM DUI 0100B

ARM

BL <target_address>

| Description | The BL (Branch with Link) instruction provides an unconditional subroutine call; the return from subroutine is achieved by copying the LR to the PC (se page 6-57). BL causes an unconditional subroutine call to a target address, and stores the return address into the LR (link register or R14). Thumb subroutine calls are made using a two-instruction sequence: |
|---|---|

1  The first instruction (H==0) sign-extends the value of <11_bit_signed_offset>, shifts the result left by 12 bits, adds the value of the PC (the address of the branch instruction plus 4), and stores the result in LR.

2  The second instruction shifts the value of <11_bit_signed_offset> left by one bit, adds the value of LR (that was calculated by the first instruction), stores the result in the PC, places the address of the instruction after the second BL in LR. *with its LSB set.*

The instruction can therefore specify a branch of +/- 4 Mbytes.

| 15 | 14 | 13 | 12 | 11 | 10 0 |
|---|---|---|---|---|---|
| 1 | 1 | 1 | 1 | H | 11_bit_signed_offset |

**Operation**

```
if H==0
        LR = PC + (SignExtend(<11_bit_signed_offset>) << 12)
else /* H==1 */
        <return_address> = (PC + 2) | 1
        PC = LR + (<11_bit_signed_offset> << 1)
        LR = <return_address>
```

*unsigned*

**Exceptions**   None

**Qualifiers**   None

**Notes**   **Memory bounds:** Branching backwards past location zero and forwards over the end of the 32-bit address space is UNPREDICTABLE.

**Offset calculation:** An assembler will calculate the branch offset address from the difference between the address of the current instruction and the address of the target (given as a program label) minus four (because the PC holds the address of the current instruction plus four).

**Equivalent ARM syntax and encoding**

BL <target_address>

| 31 | 28 | 27 | 26 | 25 | 24 | 23 | 22 | 21 | 0 |
|---|---|---|---|---|---|---|---|---|---|
| 1110 cond | | 1 | 0 | 1 | 1 | offset sign | | | 22_bit_signed_offset |

# BX

BX Rm

*Branch with exchange*

*Architecture v4T only*

**Description**   The BX (Branch and Exchange) instruction is used to branch between ARM code and Thumb code.

BX branches and selects the instruction set decoder to use to decode the instructions at the branch destination. The branch target address is the value of register Rm. The T flag is updated with bit 0 of the value of register Rm.

| 15 | 14 | 13 | 12 | 11 | 10 | 9 | 8 | 7 | 6 | 5 | 3 | 2 | 0 |
|---|---|---|---|---|---|---|---|---|---|---|---|---|---|
| 0 | 1 | 0 | 0 | 0 | 1 | 1 | 1 | 0 | H2 | Rm | | SBZ | |

**Operation**   
```
T Flag = Rm[0]
PC = Rm[31:1] << 1
```

**Exceptions**   None

**Qualifiers**   Condition Code

**Notes**   **The H2 bit:** This bit is the high register specifier:

    H2 == 0      indicates that Rm specifies a low register

    H2 == 1      indicates Rm specifies a high register

**Tranferring to Thumb:** When transferring to the Thumb instruction set, bit[0] of Rm will be set to zero when transferred to the PC.

**Transferring to ARM:** When transferring to the ARM instruction set, bit[1:0] of Rm will be set to zero when transferred to the PC.

**Equivalent ARM syntax and encoding**

BX Rm

| 31 | 30 | 29 | 28 | 27 | 26 | 25 | 24 | 23 | 22 | 21 | 20 | 19 | 16 | 15 | 12 | 11 | 8 | 7 | 6 | 5 | 4 | 3 | 0 |
|---|---|---|---|---|---|---|---|---|---|---|---|---|---|---|---|---|---|---|---|---|---|---|---|
| 1 | 1 | 1 | 0 | 0 | 0 | 0 | 1 | 0 | 0 | 1 | 0 | SBO | | SBO | | SBO | | 0 | 0 | 0 | 1 | Rm | |

**ARM Architecture Reference Manual**
ARM DUI 0100B

ARM

*Compare negative (register)*

*Architecture v4T only*

**Description**  The CMN (Compare Negative) instruction compares an arithmetic value and the negative of an arithmetic value and sets the condition code flags so that subsequent instructions can be conditionally executed (using a conditional branch instruction).

CMN performs a comparison by adding (or subtracting the negative of) the value of register Rm to (from) the value of register Rd. The condition code flags are updated (based on the result).

| 15 | 14 | 13 | 12 | 11 | 10 | 9 | 8 | 7 | 6 | 5 | | 3 | 2 | | 0 |
|----|----|----|----|----|----|---|---|---|---|---|---|---|---|---|---|
| 0 | 1 | 0 | 0 | 0 | 0 | 1 | 0 | 1 | 1 | | Rm | | | Rn | |

**Operation**
```
<alu_out> = Rn + Rm
N Flag = <alu_out>[31]
Z Flag = if <alu_out> == 0 then 1 else 0
C Flag = NOT BorrowFrom(Rn + Rm)
V Flag = OverflowFrom (Rn + Rm)
```

**Exceptions**  None

**Qualifiers**  None

## Equivalent ARM syntax and encoding

CMN Rn, Rm

| 31 | 30 | 29 | 28 | 27 | 26 | 25 | 24 | 23 | 22 | 21 | 20 | 19 | | | 16 | 15 | | | 12 | 11 | 10 | 9 | 8 | 7 | 6 | 5 | 4 | 3 | | | 0 |
|----|----|----|----|----|----|----|----|----|----|----|----|----|---|---|----|----|---|---|----|----|----|---|---|---|---|---|---|---|---|---|---|
| 1 | 1 | 1 | 0 | 0 | 0 | 0 | 1 | 0 | 1 | 1 | 1 | | Rn | | | | SBZ | | | 0 | 0 | 0 | 0 | 0 | 0 | 0 | 0 | | Rm | | |

**ARM Architecture Reference Manual**
ARM DUI 0100B

CMP Rn, #<8_bit_immediate>

*Compare immediate*

*Architecture v4T only*

**Description**  This form of the CMP (Compare) instruction compares two arithmetic values and sets the condition code flags so that subsequent instructions can be conditionally executed (using a conditional branch instruction).

In this case, CMP performs a comparison by subtracting the value of the 8-bit immediate (values 0 to 255) from the value of register Rd. The condition code flags are updated (based on the result).

| 15 | 14 | 13 | 12 | 11 | 10        8 | 7                          0 |
|----|----|----|----|----|-------------|-------------------------------|
| 0  | 0  | 1  | 0  | 1  | Rn          | 8_bit_immediate               |

**Operation**
```
<alu_out> = Rn - <8_bit_immed>
N Flag = <alu_out>[31]
Z Flag = if <alu_out> == 0 then 1 else 0
C Flag = NOT BorrowFrom(Rn - <8_bit_immed>)
V Flag = OverflowFrom (Rn - <8_bit_immed>)
```

**Exceptions**  None

**Qualifiers**  None

---

**Equivalent ARM syntax and encoding**

CMP Rn, #<8_bit_immediate>

| 31 | 30 | 29 | 28 | 27 | 26 | 25 | 24 | 23 | 22 | 21 | 20 | 19        16 | 15    12 | 11 | 10 | 9 | 8 | 7                  0 |
|----|----|----|----|----|----|----|----|----|----|----|----|--------------|----------|----|----|---|---|----------------------|
| 1  | 1  | 1  | 0  | 0  | 0  | 1  | 1  | 0  | 1  | 0  | 1  | Rn           | SBO      | 0  | 0  | 0 | 0 | 8_bit_immediate      |

**ARM Architecture Reference Manual**
ARM DUI 0100B

ARM POWERED

?

CMP Rm, Rn

**Description**

This form of the CMP (Compare) instruction compares two arithmetic values and sets the condition code flags so that subsequent instructions can be conditionally executed (using a conditional branch instruction).

In this case, CMP performs a comparison by subtracting the value of register Rm from the value of register Rd. The condition code flags are updated (based on the result).

?

Compare register

Architecture v4T only

| 15 | 14 | 13 | 12 | 11 | 10 | 9 | 8 | 7 | 6 | 5 | 3 | 2 | 0 |
|----|----|----|----|----|----|----|----|----|----|----|----|----|----|
| 0 | 1 | 0 | 0 | 0 | 0 | 1 | 0 | 1 | 0 | Rm | | Rn | |

**Operation**

```
<alu_out> = Rn - Rm
N Flag = <alu_out>[31]
Z Flag = if <alu_out> == 0 then 1 else 0
C Flag = NOT BorrowFrom(Rn - Rm)
V Flag = OverflowFrom (Rn - Rm)
```

**Exceptions**    None

**Qualifiers**    None

**Equivalent ARM syntax and encoding**

CMP Rd, Rm

| 31 | 30 | 29 | 28 | 27 | 26 | 25 | 24 | 23 | 22 | 21 | 20 | 19 | 16 | 15 | 12 | 11 | 10 | 9 | 8 | 7 | 6 | 5 | 4 | 3 | 0 |
|----|----|----|----|----|----|----|----|----|----|----|----|----|----|----|----|----|----|----|----|----|----|----|----|----|----|
| 1 | 1 | 1 | 0 | 0 | 0 | 0 | 1 | 0 | 1 | 0 | 1 | Rn | | SBZ | | 0 | 0 | 0 | 0 | 0 | 0 | 0 | 0 | Rm | |

**ARM Architecture Reference Manual**
ARM DUI 0100B

CMP Rn, Rm

**Compare
high registers**

**Architecture v4T only**

**Description**   This form of the CMP (Compare) instruction compares two arithmetic values in the high registers.

In this case, CMP :

- compares the value of a low register and a high register (H1=1, H2=0), or
- compares the value of a high register and a low register (H1=0, H2=1), or
- compares the value of a high register and another high register (H1=1, H2=1) and sets the condition code flags so that subsequent instructions can be conditionally executed (using a conditional branch instruction)

| 15 | 14 | 13 | 12 | 11 | 10 | 9 | 8 | 7 | 6 | 5 | | 3 | 2 | | 0 |
|----|----|----|----|----|----|---|---|----|----|---|---|---|---|---|---|
| 0 | 1 | 0 | 0 | 0 | 1 | 0 | 1 | H1 | H2 | Rm | | | Rd | | |

**Operation**

```
<alu_out> = Rn - Rm
N Flag = <alu_out>[31]
Z Flag = if <alu_out> == 0 then 1 else 0
C Flag = NOT BorrowFrom(Rn - Rm)
V Flag = OverflowFrom(Rn - Rm)
```

**Exceptions**   None

**Qualifiers**   None

**Notes**   **Operand restriction:**  If a low register is specified for Rd and Rm (H1=0 and H2=0), the result is UNPREDICTABLE.

**Equivalent ARM syntax and encoding**

CMP Rn, Rm

*SBZ*

| 31 | 30 | 29 | 28 | 27 | 26 | 25 | 24 | 23 | 22 | 21 | 20 | 19 | 18 | | 16 | 15 | 14 | | 12 | 11 | 10 | 9 | 8 | 7 | 6 | 5 | 4 | 3 | 2 | | 0 |
|----|----|----|----|----|----|----|----|----|----|----|----|----|----|---|----|----|----|---|----|----|----|---|---|---|---|---|---|---|---|---|---|
| 1 | 1 | 1 | 0 | 0 | 0 | 0 | 1 | 0 | 1 | 0 | 1 | H1 | | Rd | | H1 | ~~Rd~~ | | | 0 | 0 | 0 | 0 | 0 | 0 | 0 | 0 | H2 | | Rm | |

# ARM Architecture Reference Manual
ARM DUI 0100B

**Description**    The EOR (Exclusive OR) instruction can be used to invert selected bits in a register. For each bit, EOR with 1 will invert that bit, and EOR with 0 will leave it unchanged.

EOR performs a bitwise Exclusive OR of the value of register Rm with the value of register Rd, and stores the result back in register Rd. The condition code flags are updated (based on the result).

| 15 | 14 | 13 | 12 | 11 | 10 | 9 | 8 | 7 | 6 | 5 | | 3 | 2 | | 0 |
|----|----|----|----|----|----|---|---|---|---|---|---|---|---|---|---|
| 0 | 1 | 0 | 0 | 0 | 0 | 0 | 0 | 0 | 1 | | Rm | | | Rd | |

**Operation**   
```
Rd = Rd EOR Rm
N Flag = Rd[31]
Z Flag = if Rd == 0 then 1 else 0
C Flag = <shifter_carry_out>
V Flag = unaffected
```

**Exceptions**    None

**Qualifiers**    None

**Equivalent ARM syntax and encoding**

`EORS Rd, Rd, Rm`

| 31 | 30 | 29 | 28 | 27 | 26 | 25 | 24 | 23 | 22 | 21 | 20 | 19 | | 16 | 15 | | 12 | 11 | 10 | 9 | 8 | 7 | 6 | 5 | 4 | 3 | | | 0 |
|----|----|----|----|----|----|----|----|----|----|----|----|----|---|----|----|---|----|----|----|---|---|---|---|---|---|---|---|---|---|
| 1 | 1 | 1 | 0 | 0 | 0 | 0 | 0 | 0 | 0 | 1 | 1 | | Rd | | | Rd | | 0 | 0 | 0 | 0 | 0 | 0 | 0 | 0 | | Rm | | |

**ARM Architecture Reference Manual**
ARM DUI 0100B

# LDM

**LDMIA Rn!, <register_list>**

*Load multiple increment after*

*Architecture v4T only*

**Description**

The LDM instruction is useful as a block load instruction. Combined with STM (store multiple), it allows efficient block copy.

The LDMIA (Load Multiple Increment After) instruction loads a subset (or possibly all) of the general-purpose registers from sequential memory locations. The registers are loaded in sequence:

- the lowest-numbered register first, from the lowest memory address (<start_address>)

- the highest-numbered register last, from the highest memory address (<end_address>)

The <start_address> is the value of the base register Rn.

Subsequent addresses are formed by incrementing the previous address by four. One address is produced for each register that is specified in <register_list>.

The <end_address> value is four less than the sum of the value of the base register and four times the number of registers specified in <register_list>.

Finally, the base register Rn is incremented by four times the numbers of registers in <register_list>.

| 15 | 14 | 13 | 12 | 11 | 10 | 8 | 7 | 0 |
|----|----|----|----|----|----|---|---|---|
| 1 | 1 | 0 | 0 | 1 | | Rn | | register_list |

**Operation**

```
<start_address> = Rn
<end_address> = Rn + (Number_Of_Set_Bits_In(<register_list>) * 4) - 4
<address> = <start_address>
for i = 0 to 7
    if <register_list>[i] == 1
        Ri = Memory[<address>,4]
        <address> = <address> + 4
assert <end_address> == <address> - 4
Rn = Rn + (Number_Of_Set_Bits_In(<register_list>) * 4)
```

**Equivalent ARM syntax and encoding**

**LDMIA Rn!, <register_list>**

| 31 | 30 | 29 | 28 | 27 | 26 | 25 | 24 | 23 | 22 | 21 | 20 | 19 | 16 | 15 | 14 | 13 | 12 | 11 | 10 | 9 | 8 | 7 | 0 |
|----|----|----|----|----|----|----|----|----|----|----|----|----|----|----|----|----|----|----|----|---|---|---|---|
| 1 | 1 | 1 | 0 | 1 | 0 | 0 | 0 | 1 | 0 | 1 | *1 | | Rn | 0 | 0 | 0 | 0 | 0 | 0 | 0 | 0 | | register_list |

 * Writeback suppressed if Rn in list.

 **ARM Architecture Reference Manual**
ARM DUI 0100B

**Exceptions**  Data Abort

**Qualifiers**  None

**Notes**  **Register Rn:** Specifies the base register used by `<addressing_mode>`.

**Operand restrictions:** If the base register Rn is specified in `<register_list>`, the final value of Rn is the loaded value (not the written-back value).

**Data Abort:** If a data abort is signalled, the value left in Rn is IMPLEMENTATION DEFINED, but is either the original base register value or the updated base register value (even if Rn is specified in `<register_list>`).

**Non-word-aligned addresses:** Load multiple instructions ignore the least-significant two bits of `<address>` (the words are not rotated as for load word).

**Alignment:** If an implementation includes a System Control Coprocessor (see *Chapter 7, System Architecture and System Control Coprocessor*), and alignment checking is enabled, an address with bits[1:0] != 0b00 will cause an alignment exception.

`LDR Rd, [Rn, #5_bit_offset])`

**Description**   This form of the LDR (Load Register) instruction allows 32-bit memory data to be loaded into a general-purpose register where its value can be manipulated. The addressing mode is useful for accessing structure (record) fields. With an offset of zero, the address produced is the unaltered value of the base register Rn.

In this case, LDR loads a word from memory and writes it to register Rd. The memory address is calculated by adding 4 times the value of `<5_bit_offset>` to the value of register Rn. If the address is not word-aligned, the result is UNPREDICTABLE.

| 15 | 14 | 13 | 12 | 11 | 10           6 | 5        3 | 2        0 |
|----|----|----|----|----|----------------|------------|------------|
| 0  | 1  | 1  | 0  | 1  | 5_bit_offset   | Rn         | Rd         |

**Operation**
```
<address> = Rn + (5_bit_offset * 4)
if <address>[1:0] == 0b00
        <data> = Memory[<address>,4]
else
        <data> = UNPREDICTABLE
Rd = <data>
```

**Exceptions**   Data Abort

**Qualifiers**   None

**Notes**   **Alignment:** If an implementation includes a System Control Coprocessor (see *Chapter 7, System Architecture and System Control Coprocessor*), and alignment checking is enabled, an address with bits[1:0] != 0b00 will cause an alignment exception.

**Equivalent ARM syntax and encoding**

`LDR Rd, [Rn, #5_bit_offset]`

| 31 | 30 | 29 | 28 | 27 | 26 | 25 | 24 | 23 | 22 | 21 | 20 | 19    16 | 15    12 | 11 | 10 | 9 | 8 | 7 | 6       2 | 1 | 0 |
|----|----|----|----|----|----|----|----|----|----|----|----|----------|----------|----|----|---|---|---|-----------|---|---|
| 1  | 1  | 1  | 0  | 0  | 1  | 0  | 1  | 1  | 0  | 0  | 1  | Rn       | Rd       | 0  | 0  | 0 | 0 | 0 | 5_bit_offset | 0 | 0 |

**ARM Architecture Reference Manual**
ARM DUI 0100B

**Description**

This form of the LDR (Load Register) instruction allows 32-bit memory data to be loaded into a general-purpose register where its value can be manipulated. The addressing mode is useful for pointer + large offset arithmetic (use MOV immediate to set the offset), and accessing a single element of an array.

In this case, LDR loads a word from memory and writes it to register Rd. The memory address is calculated by adding the value of register Rm to the value of register Rn. If the address is not word aligned, the result is UNPREDICTABLE.

| 15 | 14 | 13 | 12 | 11 | 10 | 9 | 8 | | 6 | 5 | | 3 | 2 | | 0 |
|---|---|---|---|---|---|---|---|---|---|---|---|---|---|---|---|
| 0 | 1 | 0 | 1 | 1 | 0 | 0 | Rm | | | Rn | | | Rd | | |

**Operation**

```
<address> = Rn + Rm
if <address>[1:0] == 0b00
        <data> = Memory[<address>,4]
else
        <data> = UNPREDICTABLE
Rd = <data>
```

**Exceptions**

Data Abort

**Qualifiers**

None

**Notes**

**Alignment:** If an implementation includes a System Control Coprocessor (see *Chapter 7, System Architecture and System Control Coprocessor*), and alignment checking is enabled, an address with bits[1:0] != 0b00 will cause an alignment exception.

### Equivalent ARM syntax and encoding

LDR Rd, [Rn, Rm]

0 0 0 0 0   *Rm*

| 31 | 30 | 29 | 28 | 27 | 26 | 25 | 24 | 23 | 22 | 21 | 20 | 19 | | 16 | 15 | | 12 | 11 | 10 | 9 | 8 | 7 | 6 | | 2 | 1 | 0 |
|---|---|---|---|---|---|---|---|---|---|---|---|---|---|---|---|---|---|---|---|---|---|---|---|---|---|---|---|
| 1 | 1 | 1 | 0 | 0 | 1 | 1 | 1 | 1 | 0 | 0 | 1 | | Rn | | | Rd | | 0 | 0 | 0 | 0 | 0 | | 5-bit offset | | 0 | 0 |

LDR Rd, [PC, #8_bit_offset]

**Description**   This form of the LDR (Load Register) instruction allows 32-bit memory data to be loaded into a general-purpose register where its value can be manipulated. The addressing mode is useful for accessing PC relative data.

In this case, LDR loads a word from memory and writes it to register Rd. The memory address is calculated by adding 4 times the value of <8_bit_offset> to the value of the PC. ~~If the address is not word-aligned, the result is UNPREDICTABLE.~~ *with the two LSBits set to zero. The address is thus always word aligned*

| 15 | 14 | 13 | 12 | 11 | 10 | 8 | 7 | 0 |
|----|----|----|----|----|----|---|---|---|
| 0 | 1 | 0 | 0 | 1 | | Rd | | 8_bit_offset |

**Operation**   <address> = (PC[31:~~2~~] << ~~2~~) + (8_bit_offset * 4)
~~if <address>[1:0] == 0b00~~
         <data> = Memory[<address>,4]
~~else~~
         ~~<data> = UNPREDICTABLE~~
Rd = <data>

**Exceptions**   Data Abort

**Qualifiers**   None

**Notes**   **Alignment:** If an implementation includes a System Control Coprocessor (see *Chapter 7, System Architecture and System Control Coprocessor*), and alignment checking is enabled, an address with bits[1:0] != 0b00 will cause an alignment exception.

---

**Equivalent ARM syntax and encoding**

LDR Rd, [PC, #8_bit_offset]   *shift done by ARM*

| 31 | 30 | 29 | 28 | 27 | 26 | 25 | 24 | 23 | 22 | 21 | 20 | 19 | 18 | 17 | 16 | 15 | 12 | 11 | 10 | 9 | 2 | 1 | 0 |
|----|----|----|----|----|----|----|----|----|----|----|----|----|----|----|----|----|----|----|----|---|---|---|---|
| 1 | 1 | 1 | 0 | 0 | 1 | 0 | 1 | 1 | 0 | 0 | 1 | 1 | 1 | 1 | 1 | | Rd | 0 | 0 | | 8_bit_offset | 0 | 0 |

**ARM Architecture Reference Manual**
ARM DUI 0100B

ARM

**Description**　This form of the LDR (Load Register) instruction allows 32-bit memory data to be loaded into a general-purpose register where its value can be manipulated. The addressing mode is useful for accessing stack data.

In this case, LDR loads a word from memory and writes it to register Rd. The memory address is calculated by adding 4 times the value of `<8_bit_offset>` to the value of the SP. If the address is not word-aligned, the result is UNPREDICTABLE.

| 15 | 14 | 13 | 12 | 11 | 10 | | 8 | 7 | | | 0 |
|----|----|----|----|----|----|---|---|---|---|---|---|
| 1 | 0 | 0 | 1 | 1 | | Rd | | | 8_bit_offset | | |

**Operation**

```
<address> = SP + (8_bit_offset * 4)
if <address>[1:0] == 0b00
        <data> = Memory[<address>,4]
else
        <data> = UNPREDICTABLE
Rd = <data>
```

**Exceptions**　Data Abort

**Qualifiers**　None

**Notes**　**Alignment:** If an implementation includes a System Control Coprocessor (see *Chapter 7, System Architecture and System Control Coprocessor*), and alignment checking is enabled, an address with bits[1:0] != 0b00 will cause an alignment exception.

---

### Equivalent ARM syntax and encoding

`LDR Rd, [SP, #8_bit_offset]`

| 31 | 30 | 29 | 28 | 27 | 26 | 25 | 24 | 23 | 22 | 21 | 20 | 19 | 18 | 17 | 16 | 15 | | | | 12 | 11 | 10 | 9 | | | 2 | 1 | 0 |
|----|----|----|----|----|----|----|----|----|----|----|----|----|----|----|----|----|---|---|---|----|----|----|---|---|---|---|---|---|
| 1 | 1 | 1 | 0 | 0 | 1 | 0 | 1 | 1 | 0 | 0 | 1 | 1 | 1 | 0 | 1 | | Rd | | | | 0 | 0 | | 8_bit_offset | | | 0 | 0 |

# LDRB (1)

Load unsigned byte
immediate offset

Architecture v4T only

LDRB Rd, [Rn, #5_bit_offset]

**Description**  This form of the LDRB (Load Register Byte) instruction allows 8-bit memory data to be loaded into a general-purpose register where its value can be manipulated. The addressing mode is useful for accessing structure (record) fields.
With an offset of zero, the address produced is the unaltered value of the base register Rn.

This form of the LDRB:

1  loads a byte from memory
2  zero-extends the byte to a 32-bit word
3  writes the word to register Rd

The memory address is calculated by adding the value of <5_bit_offset> to the value of register Rn.

| 15 | 14 | 13 | 12 | 11 | 10            6 | 5        3 | 2        0 |
|----|----|----|----|----|------------------|------------|------------|
| 0  | 1  | 1  | 1  | 1  | 5_bit_offset     | Rn         | Rd         |

**Operation**
```
<address> = Rn + 5_bit_offset
     Rd = Memory[<address>,1]
```

**Exceptions**  Data Abort

**Qualifiers**  None

---

**Equivalent ARM syntax and encoding**

LDRB Rd, [Rn, #5_bit_offset]

| 31 30 29 28 | 27 26 25 24 23 22 21 20 | 19        16 | 15        12 | 11 10 9 8 7 6 | 2 1 0 |
|-------------|--------------------------|--------------|--------------|----------------|-------|
| 1 1 1 0     | 0 1 0 1 1 1 0 1          | Rn           | Rd           | 0 0 0 0 0      | 5_bit_offset | 0 0 |

**ARM Architecture Reference Manual**
ARM DUI 0100B

**Description**

This form of the LDRB (Load Register Byte) instruction allows 8-bit memory data to be loaded into a general-purpose register where its value can be manipulated. The addressing mode is useful for pointer + large offset arithmetic (use the MOV immediate to set the offset), and accessing a single element of an array.

In this case, LDRB:

1     loads a byte from memory

2     zero-extends the byte to a 32-bit word

3     writes the word to register Rd

The memory address is calculated by adding the value register Rm to the value of register Rn.

| 15 | 14 | 13 | 12 | 11 | 10 | 9 | 8 | | 6 | 5 | | 3 | 2 | | 0 |
|---|---|---|---|---|---|---|---|---|---|---|---|---|---|---|---|
| 0 | 1 | 0 | 1 | 1 | 1 | 0 | | Rm | | | Rn | | | Rd | |

**Operation**

```
<address> = Rn + Rm
Rd = Memory[<address>,1]
```

**Exceptions**     Data Abort

**Qualifiers**     None

---

### Equivalent ARM syntax and encoding

LDRB Rd, [Rn, Rm]

| 31 | 30 | 29 | 28 | 27 | 26 | 25 | 24 | 23 | 22 | 21 | 20 | 19 | | 16 | 15 | | 12 | 11 | 10 | 9 | 8 | 7 | 6 | 5 | 4 | 3 | | 1 | 0 |
|---|---|---|---|---|---|---|---|---|---|---|---|---|---|---|---|---|---|---|---|---|---|---|---|---|---|---|---|---|---|
| 1 | 1 | 1 | 0 | 0 | 1 | 1 | 1 | 1 | 1 | 0 | 1 | | Rn | | | Rd | | 0 | 0 | 0 | 0 | 0 | 0 | 0 | 0 | | Rm | | |

**ARM Architecture Reference Manual**
ARM DUI 0100B

# LDRH (1)

LDRH Rd, [Rn, #5_bit_offset]

Load unsigned halfword
Immediate offset

Architecture v4T only

**Description** This form of the LDRH (Load Register Halfword) instruction allows 16-bit memory data to be loaded into a general-purpose register where its value can be manipulated. The addressing mode is useful for accessing structure (record) fields. With an offset of zero, the address produced is the unaltered value of the base register Rn.

In this case, LDRH:

1 loads a halfword from memory

2 zero-extends the halfword to a 32-bit word

3 writes the word to register Rd

The memory address is calculated by adding 2 times the value of <5_bit_offset> to the value of register Rn. If the address is not halfword-aligned, the result is UNPREDICTABLE.

| 15 | 14 | 13 | 12 | 11 | 10　　　　　　　6 | 5　　　3 | 2　　　0 |
|----|----|----|----|----|----|----|----|
| 1 | 0 | 0 | 0 | 1 | 5_bit_offset | Rn | Rd |

**Operation**
```
<address> = Rn + (5_bit_offset * 2)
if <address>[X:0] == 0
        <data> = Memory[<address>,2]
else
        <data> = UNPREDICTABLE
Rd = <data>
```

**Exceptions** Data Abort

**Qualifiers** None

**Notes** **Alignment:** If an implementation includes a System Control Coprocessor (see *Chapter 7, System Architecture and System Control Coprocessor*), and alignment checking is enabled, an address with bit[0] != 0 will cause an alignment exception.

---

**Equivalent ARM syntax and encoding**

LDRH Rd, [Rn, #5_bit_offset]

| 31 | 30 | 29 | 28 | 27 | 26 | 25 | 24 | 23 | 22 | 21 | 20 | 19　　　16 | 15　　　12 | 11 | 10 | 9 | 8 | 7 | 6 | 54 | 3　　　0 |
|----|----|----|----|----|----|----|----|----|----|----|----|----|----|----|----|----|----|----|----|----|----|
| 1 | 1 | 1 | 0 | 0 | 0 | 0 | 1 | 1 | 1 | 0 | 1 | Rn | Rd | 0 | 0 | 0 | O4 | 1 | 0 | 1 1 | offset[3:0] |

---

**ARM Architecture Reference Manual**
ARM DUI 0100B

**Description**  This form of the LDRH (Load Register Halfword) instruction allows 16-bit memory data to be loaded into a general-purpose register where its value can be manipulated. The addressing mode is useful for pointer + large offset arithmetic (use MOV immediate to set the offset), and accessing a single element of an array.

In this case, LDRH:

1   loads a halfword from memory

2   zero-extends the halfword to a 32-bit word

3   writes the word to register Rd

The memory address is calculated by adding the value of register Rm to the value of register Rn. If the address is not halfword-aligned, the result is UNPREDICTABLE.

| 15 | 14 | 13 | 12 | 11 | 10 | 9 | 8 | | 6 | 5 | | 3 | 2 | | 0 |
|----|----|----|----|----|----|---|---|---|---|---|---|---|---|---|---|
| 0 | 1 | 0 | 1 | 1 | 0 | 1 | | Rm | | | Rn | | | Rd | |

**Operation**
```
<address> = Rn + Rm
if <address>[0] == 0
        <data> = Memory[<address>,2]
else
        <data> = UNPREDICTABLE
Rd = <data>
```

**Exceptions**  Data Abort

**Qualifiers**  None

**Notes**  **Alignment:** If an implementation includes a System Control Coprocessor (see *Chapter 7, System Architecture and System Control Coprocessor*), and alignment checking is enabled, an address with bit[0] != 0 will cause an alignment exception.

---

### Equivalent ARM syntax and encoding

`LDRH Rd, [Rn, Rm]`

| 31 | 30 | 29 | 28 | 27 | 26 | 25 | 24 | 23 | 22 | 21 | 20 | 19 | | 16 | 15 | | 12 | 11 | | 8 | 7 | 6 | 5 | 4 | 3 | | 0 |
|----|----|----|----|----|----|----|----|----|----|----|----|----|---|----|----|---|----|----|---|---|---|---|---|---|---|---|---|
| 1 | 1 | 1 | 0 | 0 | 0 | 0 | 1 | 1 | 0 | 0 | 1 | | Rn | | | Rd | | | SBZ | | 1 | 0 | 1 | 1 | | Rm | |

**ARM Architecture Reference Manual**
ARM DUI 0100B

# LDRSB

`LDRSB Rd, [Rn, Rm]`

**Description**     The LDRSB (Load Register Signed Byte) instruction allows 8-bit signed memory data to be loaded into a general-purpose register where its value can be manipulated. The addressing mode is useful for pointer + large offset arithmetic (use the MOV immediate to set the offset), and accessing a single element of an array.

LDRSB:

1   loads a byte from memory

2   sign-extends the byte to a 32-bit word

3   writes it to register Rd

The memory address is calculated by adding the value register Rm to the value of register Rn.

| 15 | 14 | 13 | 12 | 11 | 10 | 9 | 8 | 6 | 5 | 3 | 2 | 0 |
|----|----|----|----|----|----|---|---|---|---|---|---|---|
| 0 | 1 | 0 | 1 | 0 | 1 | 1 | Rm | | Rn | | Rd | |

**Operation**     `<address> = Rn + Rm`
`Rd = SignExtend(Memory[<address>,1])`

**Exceptions**    Data Abort

**Qualifiers**    None

---

### Equivalent ARM syntax and encoding

`LDRSB Rd, [Rn, Rm]`

| 31 | 30 | 29 | 28 | 27 | 26 | 25 | 24 | 23 | 22 | 21 | 20 | 19 | 16 | 15 | 12 | 11 | 8 | 7 | 6 | 5 | 4 | 3 | 0 |
|----|----|----|----|----|----|----|----|----|----|----|----|----|----|----|----|----|---|---|---|---|---|---|---|
| 1 | 1 | 1 | 0 | 0 | 0 | 0 | 1 | 1 | 0 | 0 | 1 | Rn | | Rd | | SBZ | | 1 | 1 | 0 | 1 | Rm | |

**ARM Architecture Reference Manual**
ARM DUI 0100B

ARM

**Description**

The LDRSH (Load Register Signed Halfword) instruction allows 16-bit signed memory data to be loaded into a general-purpose register where its value can be manipulated. The addressing mode is useful for pointer + large offset arithmetic (use the MOV immediate to set the offset), and accessing a single element of an array.

LDRSH:

1  loads a halfword from memory

2  sign-extends the halfword to a 32-bit word

3  writes the word to register Rd

The memory address is calculated by adding the value register Rm to the value of register Rn. If the address is not halfword-aligned, the result is UNPREDICTABLE.

| 15 | 14 | 13 | 12 | 11 | 10 | 9 | 8 | | 6 | 5 | | 3 | 2 | | 0 |
|----|----|----|----|----|----|---|---|---|---|---|---|---|---|---|---|
| 0 | 1 | 0 | 1 | 1 | 1 | 1 | | Rm | | | Rn | | | Rd | |

**Operation**

```
<address> = Rn + Rm
if <address>[X:0] == 0
        <data> = Memory[<address>,2]
else
        <data> = UNPREDICTABLE
Rd = SignExtend(<data>)
```

**Exceptions**    Data Abort

**Qualifiers**    None

**Notes**    **Alignment:** If an implementation includes a System Control Coprocessor (see *Chapter 7, System Architecture and System Control Coprocessor*), and alignment checking is enabled, an address with bit[0] != 0 will cause an alignment exception.

**Equivalent ARM syntax and encoding**

LDRSH Rd, [Rn, Rm

| 31 | 30 | 29 | 28 | 27 | 26 | 25 | 24 | 23 | 22 | 21 | 20 | 19 | | 16 | 15 | | 12 | 11 | | 8 | 7 | 6 | 5 | 4 | 3 | | 0 |
|----|----|----|----|----|----|----|----|----|----|----|----|----|---|----|----|---|----|----|---|---|---|---|---|---|---|---|---|
| 1 | 1 | 1 | 0 | 0 | 0 | 0 | 1 | 1 | 0 | 0 | 1 | | Rn | | | Rd | | | SBZ | | 1 | 1 | 1 | 1 | | Rm | |

**ARM Architecture Reference Manual**
ARM DUI 0100B

## LSL (1)

*Logical shift left (immediate)*

*Architecture v4T only*

LSL Rd, Rm, #<shift_imm>

**Description**  This form of the LSL (Logical Shift Left) instruction is used to provide either the value of a register directly (LSL #0), or the value of a register multiplied by a constant power of two.

In this case, LSL performs a logical shift left of the value of register Rm by an immediate value in the range 0 to 31 and stores the result into the destination register Rd. Zeros are inserted into the vacated bit positions. The condition code flags are updated (based on the result).

| 15 | 14 | 13 | 12 | 11 | 10 | 6 | 5 | 3 | 2 | 0 |
|----|----|----|----|----|----|----|----|----|----|----|
| 0 | 0 | 0 | 0 | 0 | shift_imm | | Rm | | Rd | |

**Operation**
```
if <shift_imm> == 0
        C Flag = UNAFFECTED
        Rd = UNAFFECTED
else /* <shift_imm> > 0 */
        C Flag = Rm[32 - <shift_imm>]
        Rd = Rm Logical_Shift_Left <shift_imm>
N Flag = Rd[31]
Z Flag = if Rd == 0 then 1 else 0
V Flag = unaffected
```

**Exceptions**  None

**Qualifiers**  None

**Equivalent ARM syntax and encoding**

MOVS Rd, Rm, LSL #<shift_imm>

| 31 | 30 | 29 | 28 | 27 | 26 | 25 | 24 | 23 | 22 | 21 | 20 | 19 | 16 | 15 | 12 | 11 | 7 | 6 | 5 | 4 | 3 | 0 |
|----|----|----|----|----|----|----|----|----|----|----|----|----|----|----|----|----|----|----|----|----|----|----|
| 1 | 1 | 1 | 0 | 0 | 0 | 0 | 1 | 1 | 0 | 1 | 1 | SBZ | | Rd | | shift_imm | | 0 | 0 | 0 | Rm | |

**ARM Architecture Reference Manual**
ARM DUI 0100B

This form of the LSL (Logical Shift Left) instruction is used to provide the unsigned value of a register multiplied by a variable (in a register) power of two.

In this case, LSL instruction performs a logical shift left of the value of register Rd by the value in the least-significant byte of register Rs and stores the result back into the register Rd. Zeros are inserted into the vacated bit positions. The condition code flags are updated (based on the result).

| 15 | 14 | 13 | 12 | 11 | 10 | 9 | 8 | 7 | 6 | 5 | | 3 | 2 | | 0 |
|----|----|----|----|----|----|---|---|---|---|---|---|---|---|---|---|
| 0 | 1 | 0 | 0 | 0 | 0 | 0 | 0 | 1 | 0 | | Rs | | | Rd | |

**Operation**

```
if Rs[7:0] == 0
        C Flag = UNAFFECTED
        Rd = UNAFFECTED
else if Rs[7:0] < 32 then
        C Flag = Rd[32 - Rs[7:0]]
        Rd = Rd Logical_Shift_Left Rs[7:0]
else if Rs[7:0] == 32 then
        C Flag = Rd[0]
        Rd = 0
else /* Rs[7:0] > 32 */
        C Flag = 0
        Rd = 0
N Flag = Rd[31]
Z Flag = if Rd == 0 then 1 else 0
V Flag = unaffected
```

**Exceptions**    None

**Qualifiers**    None

### Equivalent ARM syntax and encoding

```
MOVS Rd, Rd, LSL Rs
```

| 31 | 30 | 29 | 28 | 27 | 26 | 25 | 24 | 23 | 22 | 21 | 20 | 19 | 16 | 15 | 12 | 11 | 8 | 7 | 6 | 5 | 4 | 3 | 0 |
|----|----|----|----|----|----|----|----|----|----|----|----|----|----|----|----|----|---|---|---|---|---|---|---|
| 1 | 1 | 1 | 0 | 0 | 0 | 0 | 1 | 1 | 0 | 1 | 1 | SBZ | | Rd | | Rs | | 0 | 0 | 0 | 1 | Rd | |

**ARM Architecture Reference Manual**
ARM DUI 0100B

# LSR (1)

Logical shift right
(immediate)

Architecture v4T only

`LSR Rd, Rm, #<shift_imm>`

**Description**    This form of the LSR (Logical Shift Right) instruction is used to provide the value of a register, divided by a constant power of two.

In this case, LSR performs a logical shift right of the value of register Rm by an immediate value in the range 1 to 32, and stores the result into the destination register Rd. Zeros are inserted into the vacated bit positions.

A shift by 32 is encoded by:

`<shift_imm> = 0`

The condition code flags are updated (based on the result).

| 15 | 14 | 13 | 12 | 11 | 10 | | | | | 6 | 5 | | 3 | 2 | | | 0 |
|----|----|----|----|----|----|---|---|---|---|---|---|---|---|---|---|---|---|
| 0 | 0 | 0 | 0 | 1 | shift_imm | | | | | | Rm | | | Rd | | | |

**Operation**
```
if <shift_imm> == 0
        C Flag = Rm[31]
        Rd = 0
else /* <shift_imm> > 0 */
        C Flag = Rm[<shift_imm> - 1]
        Rd = Rm Logical_Shift_Right <shift_imm>
N Flag = Rd[31]
Z Flag = if Rd == 0 then 1 else 0
V Flag = unaffected
```

**Exceptions**    None

**Qualifiers**    None

---

**Equivalent ARM syntax and encoding**

`MOVS Rd, Rm, LSR #<shift_imm>`

| 31 | 30 | 29 | 28 | 27 | 26 | 25 | 24 | 23 | 22 | 21 | 20 | 19 | | 16 | 15 | | 12 | 11 | | | 7 | 6 | 5 | 4 | 3 | | | 0 |
|----|----|----|----|----|----|----|----|----|----|----|----|----|---|----|----|---|----|----|---|---|---|---|---|---|---|---|---|---|
| 1 | 1 | 1 | 0 | 0 | 0 | 0 | 1 | 1 | 0 | 1 | 1 | SBZ | | | Rd | | | shift_imm | | | 0 | 1 | 0 | | Rm | | | |

**ARM Architecture Reference Manual**
ARM DUI 0100B

ARM

`LSR Rd, Rs`

**Description**  This form of the LSR (Logical Shift Right) instruction is used to provide the unsigned value of a register divided by a variable (in a register) power of two.

In this case, LSR instruction performs a logical shift right of the value of register Rd by the value in the least-significant byte of register Rs, and stores the result back into the register Rd. Zeros are inserted into the vacated bit positions. The condition code flags are updated (based on the result).

| 15 | 14 | 13 | 12 | 11 | 10 | 9 | 8 | 7 | 6 | 5 | | 3 | 2 | | 0 |
|----|----|----|----|----|----|---|---|---|---|---|---|---|---|---|---|
| 0 | 1 | 0 | 0 | 0 | 0 | 0 | 0 | 1 | 1 | | Rs | | | Rd | |

**Operation**
```
if Rs[7:0] == 0 then
        C Flag = unaffected
        Rd = unaffected
else if Rs[7:0] < 32 then
        C Flag = Rd[Rs[7:0] - 1]
        Rd = Rd Logical_Shift_Right Rs[7:0]
else if Rs[7:0] == 32 then
        C Flag = Rd[31]
        Rd = 0
else /* Rs[7:0] > 32 */
        C Flag = 0
        Rd = 0
N Flag = Rd[31]
Z Flag = if Rd == 0 then 1 else 0
V Flag = unaffected
```

**Exceptions**  None

**Qualifiers**  None

### Equivalent ARM syntax and encoding

`MOVS Rd, Rd, LSR Rs`

| 31 | 30 | 29 | 28 | 27 | 26 | 25 | 24 | 23 | 22 | 21 | 20 | 19 | | 16 | 15 | | 12 | 11 | | 8 | 7 | 6 | 5 | 4 | 3 | | 0 |
|----|----|----|----|----|----|----|----|----|----|----|----|----|---|----|----|---|----|----|---|---|---|---|---|---|---|---|---|
| 1 | 1 | 1 | 0 | 0 | 0 | 0 | 1 | 1 | 0 | 1 | 1 | | SBZ | | | Rd | | | Rs | | 0 | 0 | 1 | 1 | | Rd | |

**ARM Architecture Reference Manual**
ARM DUI 0100B

# MOV (1)

*Move immediate*

*Architecture v4T only*

MOV Rd, #<8_bit_immediate>

**Description**   This form of the MOV (Move) instruction moves a large constant value to a register.

In this case, MOV writes the value of the 8-bit immediate (values 0 to 255) to the destination register Rd. The condition code flags are updated (based on the result).

| 15 | 14 | 13 | 12 | 11 | 10 | 8 | 7 | 0 |
|----|----|----|----|----|----|---|---|---|
| 0 | 0 | 1 | 0 | 0 | | Rd | | 8_bit_immediate |

**Operation**
```
Rd = <8_bit_immed>
N Flag = Rd[31]
Z Flag = if Rd == 0 then 1 else 0
C Flag = unaffected
V Flag = unaffected
```

**Exceptions**   None

**Qualifiers**   None

**Equivalent ARM syntax and encoding**

MOV Rn, #<8_bit_immediate>

| 31 | 30 | 29 | 28 | 27 | 26 | 25 | 24 | 23 | 22 | 21 | 20 | 19 | 16 | 15 | 12 | 11 | 10 | 9 | 8 | 7 | 0 |
|----|----|----|----|----|----|----|----|----|----|----|----|----|----|----|----|----|----|---|---|---|---|
| 1 | 1 | 1 | 0 | 0 | 0 | 1 | 1 | 1 | 0 | 0 | 1 | | SBZ | | Rd | 0 | 0 | 0 | 0 | | 8_bit_immediate |

**ARM Architecture Reference Manual**

ARM DUI 0100B

**Description**    This form of the MOV (Move) instruction is used to move a value to, from, or between high registers.

In this case, MOV :

- moves the value of a low register to a high register (H1=1, H2=0), or
- moves the value of a high register to a low register (H1=0, H2=1), or
- moves the value of a high register and another high register (H1=1, H2=1).

The subroutine return instruction is:

```
MOV PC, LR
```

(after executing a BL sequence; see page 6-33).

| 15 | 14 | 13 | 12 | 11 | 10 | 9 | 8 | 7 | 6 | 5 | | 3 | 2 | | 0 |
|----|----|----|----|----|----|---|---|----|----|---|---|---|---|---|---|
| 0 | 1 | 0 | 0 | 0 | 1 | 1 | 0 | H1 | H2 | | Rm | | | Rd | |

**Operation**    Rd = Rm

**Exceptions**    None

**Qualifiers**    None

**Notes**    **Operand restriction:** If a low register is specified for Rd and Rm (H1=0 and H2=0), the result is UNPREDICTABLE.

---

### Equivalent ARM syntax and encoding

```
MOV Rd, Rm
```

*SBZ*

| 31 | 30 | 29 | 28 | 27 | 26 | 25 | 24 | 23 | 22 | 21 | 20 | 19 | 18 | | 16 | 15 | 14 | | 12 | 11 | 10 | 9 | 8 | 7 | 6 | 5 | 4 | 3 | 2 | | 0 |
|----|----|----|----|----|----|----|----|----|----|----|----|----|----|---|----|----|----|---|----|----|----|---|---|---|---|---|---|---|---|---|---|
| 1 | 1 | 1 | 0 | 0 | 0 | 0 | 1 | 1 | 0 | 1 | 0 | H1 | | Rd | | H1 | Rd | | | 0 | 0 | 0 | 0 | 0 | 0 | 0 | 0 | 0 | H2 | | Rm | |

# MUL

*Multiply*

*Architecture v4T only*

**MUL Rd, Rm**

**Description**

The MUL (Multiply) instruction multiplies signed or unsigned variables to produce a 32-bit result.

MUL multiplies the value of register Rm with the value of register Rd, and stores the result back in the register Rd. The condition code flags are updated (based on the result).

| 15 | 14 | 13 | 12 | 11 | 10 | 9 | 8 | 7 | 6 | 5 | | | 3 | 2 | | | 0 |
|----|----|----|----|----|----|----|----|----|----|----|----|----|----|----|----|----|----|
| 0 | 1 | 0 | 0 | 0 | 0 | 1 | 1 | 0 | 1 | Rm | | | | Rd | | | |

**Operation**

```
Rd = (Rm * Rd)[31:0]
N Flag = Rd[31]
Z Flag = if Rd == 0 then 1 else 0
C Flag = UNPREDICTABLE
V Flag = UNAFFECTED
```

**Exceptions**     None

**Qualifiers**     None

**Notes**

**Operand restriction:** Specifying the same register for Rd and Rm has UNPREDICTABLE results.

**Early termination:** If the multiplier implementation supports early termination, it must be implemented on the value of the Rd operand. The type of early termination used (signed or unsigned) is IMPLEMENTATION DEFINED.

**Signed and unsigned:** Because the MUL instruction produces only the lower 32 bits of the 64-bit product, MUL gives the same answer for multiplication of both signed and unsigned numbers.

---

**Equivalent ARM syntax and encoding**

**MULS Rd, Rm, Rd**

| 31 | 30 | 29 | 28 | 27 | 26 | 25 | 24 | 23 | 22 | 21 | 20 | 19 | | | 16 | 15 | | | 12 | 11 | | | 8 | 7 | 6 | 5 | 4 | 3 | | | 0 |
|----|----|----|----|----|----|----|----|----|----|----|----|----|----|----|----|----|----|----|----|----|----|----|----|----|----|----|----|----|----|----|----|
| 1 | 1 | 1 | 0 | 0 | 0 | 0 | 0 | 0 | 0 | 0 | 1 | Rd | | | | SBZ | | | | Rd | | | | 1 | 0 | 0 | 1 | Rm | | | |

**ARM Architecture Reference Manual**

ARM DUI 0100B

| Description | The MVN (Move NOT) instruction is used to compliment a register value, often to form a bit mask. |
|---|---|
| | MVN writes the logical one's compliment value of register Rn to the destination register Rd. The condition code flags are updated (based on the result). |

*Architecture v4T only*

| 15 | 14 | 13 | 12 | 11 | 10 | 9 | 8 | 7 | 6 | 5 | 3 | 2 | 0 |
|----|----|----|----|----|----|---|---|---|---|---|---|---|---|
| 0 | 0 | 0 | 0 | 0 | 0 | 1 | 1 | 1 | 1 | Rm | | Rd | |

| Operation | ```
Rd = NOT Rm
N Flag = Rd[31]
Z Flag = if Rd == 0 then 1 else 0
C Flag = unaffected
V Flag = unaffected
``` |
|---|---|
| **Exceptions** | None |
| **Qualifiers** | None |

### Equivalent ARM syntax and encoding

`MVNS Rd, Rm`

| 31 | 30 | 29 | 28 | 27 | 26 | 25 | 24 | 23 | 22 | 21 | 20 | 19 | 16 | 15 | 12 | 11 | 10 | 9 | 8 | 7 | 6 | 5 | 4 | 3 | 0 |
|----|----|----|----|----|----|----|----|----|----|----|----|----|----|----|----|----|----|---|---|---|---|---|---|---|---|
| 1 | 1 | 1 | 0 | 0 | 0 | 0 | 1 | 1 | 1 | 1 | 1 | SBZ | | Rd | | 0 | 0 | 0 | 0 | 0 | 0 | 0 | 0 | Rm | |
```

# NEG

*Negate register*

*Architecture v4T only*

NEG Rd, Rn

**Description**   The NEG (Negate) instruction negates the value of one register and stores the result in a second register.

NEG subtracts the value of register Rn from zero, and stores the result in the destination register Rd. The condition code flags are updated (based on the result).

| 15 | 14 | 13 | 12 | 11 | 10 | 9 | | | 6 | 5 | | 3 | 2 | | 0 |
|---|---|---|---|---|---|---|---|---|---|---|---|---|---|---|---|
| 0 | 1 | 0 | 0 | 0 | 0 | 1 | 0 | 0 | 1 | | Rn | | | Rd | |

**Operation**
```
Rd = 0 - Rn
N Flag = Rd[31]
Z Flag = if Rd == 0 then 1 else 0
C Flag = NOT BorrowFrom(0 - Rn)
V Flag = OverflowFrom(0 - Rn)
```

**Exceptions**   None

**Qualifiers**   None

**Equivalent ARM syntax and encoding**

RSBS Rd, Rn, #0

| 31 | 30 | 29 | 28 | 27 | 26 | 25 | 24 | 23 | 22 | 21 | 20 | 19 | | | 16 | 15 | | | 12 | 11 | 10 | 9 | 8 | 7 | 6 | 5 | 4 | 3 | 2 | 1 | 0 |
|---|---|---|---|---|---|---|---|---|---|---|---|---|---|---|---|---|---|---|---|---|---|---|---|---|---|---|---|---|---|---|---|
| 1 | 1 | 1 | 0 | 0 | 0 | 1 | 0 | 0 | 1 | 1 | 1 | | Rn | | | | Rd | | | 0 | 0 | 0 | 0 | 0 | 0 | 0 | 0 | 0 | 0 | 0 | 0 |

**ARM Architecture Reference Manual**
ARM DUI 0100B

**Description**  The ORR (Logical OR) instruction can be used to set selected bits in a register; for each bit, ORR with 1 will set the bit, and ORR with 0 will leave it unchanged.

ORR performs a bitwise (inclusive) OR of the value of register Rm with the value of register Rd, and stores the result back in register Rd. The condition code flags are updated (based on the result).

| 15 | 14 | 13 | 12 | 11 | 10 | 9 | 8 | 7 | 6 | 5 | | 3 | 2 | | 0 |
|----|----|----|----|----|----|---|---|---|---|---|---|---|---|---|---|
| 0 | 1 | 0 | 0 | 0 | 0 | 1 | 1 | 0 | 0 | | Rm | | | Rd | |

**Operation**
```
Rd = Rd OR Rm
N Flag = Rd[31]
Z Flag = if Rd == 0 then 1 else 0
C Flag = <shifter_carry_out>
V Flag = unaffected
```

**Exceptions**  None

**Qualifiers**  None

---

**Equivalent ARM syntax and encoding**

ORRS `Rd, Rd, Rm`

| 31 | 30 | 29 | 28 | 27 | 26 | 25 | 24 | 23 | 22 | 21 | 20 | 19 | | 16 | 15 | | 12 | 11 | 10 | 9 | 8 | 7 | 6 | 5 | 4 | 3 | | | 0 |
|----|----|----|----|----|----|----|----|----|----|----|----|----|---|----|----|---|----|----|----|---|---|---|---|---|---|---|---|---|---|
| 1 | 1 | 1 | 0 | 0 | 0 | 0 | 1 | 1 | 0 | 0 | 1 | | Rd | | | Rd | | 0 | 0 | 0 | 0 | 0 | 0 | 0 | 0 | | Rm | | |

POP {<register_list>, <PC>}

**Description** The POP (Pop Multiple Registers) instruction is useful for stack operations, including procedure exit, to restore saved registers, load the PC with the return address, and update the stack pointer.

POP loads a subset (or possibly all) of the general-purpose registers and optionally the PC from sequential memory locations. Registers are loaded in sequence:

- the lowest-numbered register first, from the lowest memory address (<start_address>)

- the highest-numbered register last, from the highest memory address (<end_address>)

The <start_address> is the value of the SP.

Subsequent addresses are formed by incrementing the previous address by four. One address is produced for each register that is specified in <register_list>.

The <end_address> value is four less than the sum of the value of the SP and four times the number of registers specified in <register_list> (including the R bit).

Finally, the base register Rn is incremented by four times the numbers of registers in <register_list> (plus the R bit).

| 15 | 14 | 13 | 12 | 11 | 10 | 9 | 8 | 7 | | 0 |
|----|----|----|----|----|----|---|---|---|---|---|
| 1 | 0 | 1 | 1 | 1 | 1 | 0 | R | | register_list | |

**Operation**

```
<start_address> = Rn
<end_address> = Rn + (Number_Of_Set_Bits_In(<register_list> + R) * 4) - 4
<address> = <start_address>
for i = 0 to 7
    if <register_list>[i] == 1
        Ri = Memory[<address>,4]
        <address> = <address> + 4
    if R == 1
        PC = Memory[<address>,4]
        <address> = <address> + 4
assert <end_address> == <address> - 4
Rn = Rn + (Number_Of_Set_Bits_In(<register_list> + R) * 4)
```

**Equivalent ARM syntax and encoding**

LDMIA SP!, <register_list>, {PC}

| 31 | 30 | 29 | 28 | 27 | 26 | 25 | 24 | 23 | 22 | 21 | 20 | 19 | 18 | 17 | 16 | 15 | 14 | 13 | 12 | 11 | 10 | 9 | 8 | 7 | | 0 |
|----|----|----|----|----|----|----|----|----|----|----|----|----|----|----|----|----|----|----|----|----|----|---|---|---|---|---|
| 1 | 1 | 1 | 0 | 1 | 0 | 0 | 0 | 1 | 0 | 1 | 1 | 1 | 1 | 0 | 1 | R | 0 | 0 | 0 | 0 | 0 | 0 | 0 | | register_list | |

**ARM Architecture Reference Manual**

ARM DUI 0100B

**Exceptions**     Data Abort

**Qualifiers**     None

**Notes**     **The R bit:** If R == 1, the PC is also loaded from the stack; if R == 0, the PC is not loaded.

**Data Abort:** If a data abort is signalled, the value left in SP is IMPLEMENTATION DEFINED, but is either the original SP value or the updated SP value.

**Non-word aligned addresses:** Pop multiple instructions ignore the least-significant two bits of <address> (the words are not rotated as for load word).

**Alignment:** If an implementation includes a System Control Coprocessor (see *Chapter 7, System Architecture and System Control Coprocessor*), and alignment checking is enabled, an address with bits[1:0] != 0b00 will cause an alignment exception.

PUSH {<register_list>, <LR>}

The PUSH (Push Multiple Registers) instruction is useful for stack operations, including procedure entry, to save registers (optionally including the return address), and to update the stack pointer.

PUSH stores a subset (or possibly all) of the general-purpose registers and optionally the LR to sequential memory locations. The registers are stored in sequence:

- the lowest-numbered register first, to the lowest memory address (<start_address>)
- the highest-numbered register last, to the highest memory address (<end_address>)

The <start_address> is the value of the SP.

Subsequent addresses are formed by incrementing the previous address by four. One address is produced for each register that is specified in <register_list>.

The <end_address> value is four less than the sum of the value of the SP and four times the number of registers specified in <register_list> (including the R bit).

Finally, the base register Rn is incremented by four times the numbers of registers in <register_list> (plus the R bit).

| 15 | 14 | 13 | 12 | 11 | 10 | 9 | 8 | 7 | | 0 |
|---|---|---|---|---|---|---|---|---|---|---|
| 1 | 0 | 1 | 1 | 0 | 1 | 0 | R | | register_list | |

### Operation

```
<start_address> = Rn
<end_address> = Rn + (Number_Of_Set_Bits_In(<register_list> + R) * 4) - 4
<address> = <start_address>
for i = 0 to 7
    if <register_list>[i] == 1
        Memory[<address>,4] = Ri
        <address> = <address> + 4
    if R == 1
        Memory[<address>,4] = LR
        <address> = <address> + 4
assert <end_address> == <address> - 4
Rn = Rn + (Number_Of_Set_Bits_In(<register_list> + R) * 4)
```

### Equivalent ARM syntax and encoding

$ST$

~~LDMDB~~ SP!, <register_list>, {LR}

| 31 | 30 | 29 | 28 | 27 | 26 | 25 | 24 | 23 | 22 | 21 | 20 | 19 | 18 | 17 | 16 | 15 | 14 | 13 | 12 | 11 | 10 | 9 | 8 | 7 | | 0 |
|---|---|---|---|---|---|---|---|---|---|---|---|---|---|---|---|---|---|---|---|---|---|---|---|---|---|---|
| 1 | 1 | 1 | 0 | 1 | 0 | 0 | 1 | 0 | 0 | 1 | 0 | 1 | 1 | 0 | 1 | 0 | R | 0 | 0 | 0 | 0 | 0 | 0 | | register_list | |

## ARM Architecture Reference Manual

ARM DUI 0100B

| | |
|---|---|
| **Exceptions** | Data Abort |
| **Qualifiers** | None |

**Notes**

**The R bit:** If R == 1, the LR is also stored to the stack; if R == 0, the LR is not stored.

**Data Abort:** If a data abort is signalled, the value left in SP is IMPLEMENTATION DEFINED, but is either the original SP value or the updated SP value.

**Non-word aligned addresses:** Push multiple instructions ignore the least-significant two bits of `<address>` (the words are not rotated as for load word).

**Alignment:** If an implementation includes a System Control Coprocessor (see *Chapter 7, System Architecture and System Control Coprocessor*), and alignment checking is enabled, an address with bits[1:0] != 0b00 will cause an alignment exception.

**ARM Architecture Reference Manual**
ARM DUI 0100B

ROR Rd, Rs

**Description**  The ROR (Rotate Right Register) instruction is used to provide the value of a register rotated by a variable value (in a register).

ROR performs a rotate right of the value of register Rd by the value in the least-significant byte of register Rs, and stores the result back into register Rd. The bits that are rotated off the right end are inserted into the vacated bit positions on the left. The condition code flags are updated (based on the result).

| 15 | 14 | 13 | 12 | 11 | 10 | 9 | 8 | 7 | 6 | 5 | | 3 | 2 | | 0 |
|----|----|----|----|----|----|---|---|---|---|---|---|---|---|---|---|
| 0 | 1 | 0 | 0 | 0 | 0 | 0 | 1 | 1 | 1 | | Rs | | | Rd | |

**Operation**
```
if Rs[7:0] == 0 then
        C Flag = unaffected
        Rd = unaffected
else if Rs[4:0] == 0 then
        C Flag = Rd[31]
        Rd = unaffected
else /* Rs[4:0] > 0 */
        C Flag = Rd[Rs[4:0] - 1]
        Rd = Rd Rotate_Right Rs[4:0]
```

**Exceptions**  None

**Qualifiers**  None

### Equivalent ARM syntax and encoding

MOVS Rd, Rd, ROR Rs

| 31 | 30 | 29 | 28 | 27 | 26 | 25 | 24 | 23 | 22 | 21 | 20 | 19 | | 16 | 15 | | 12 | 11 | | 8 | 7 | 6 | 5 | 4 | 3 | | | 0 |
|----|----|----|----|----|----|----|----|----|----|----|----|----|---|----|----|---|----|----|---|---|---|---|---|---|---|---|---|---|
| 1 | 1 | 1 | 0 | 0 | 0 | 0 | 1 | 1 | 0 | 1 | 1 | | SBZ | | | Rd | | | Rs | | 0 | 1 | 1 | 1 | | Rd | | |

**ARM Architecture Reference Manual**
ARM DUI 0100B

**Description**   The SBC (Subtract with Carry) instruction is used to synthesize 64-bit subtraction. If register pairs R0,R1 and R2,R3 hold 64-bit values (R0 and R2 hold the least-significant word), the following instructions leave the 64-bit sum in R0,R1.

```
SUB     R0,R2
SBC     R1,R3
```

SBC subtracts the value of register Rm and the value of NOT(Carry Flag) from the value of register Rd, and stores the result in register Rd. The condition code flags are updated (based on the result).

| 15 | 14 | 13 | 12 | 11 | 10 | 9 | | 6 | 5 | | 3 | 2 | | 0 |
|----|----|----|----|----|----|---|---|---|---|---|---|---|---|---|
| 0 | 1 | 0 | 0 | 0 | 0 | 0 | 1 | 1 | 0 | Rm | | Rd | | |

**Operation**
```
Rd = Rd - Rm - NOT(C Flag)
N Flag = Rd[31]
Z Flag = if Rd == 0 then 1 else 0
C Flag = CarryFrom(Rd - Rm - NOT(C Flag))
V Flag = OverflowFrom(Rd - Rm - NOT(C Flag))
```

**Exceptions**   None

**Qualifiers**   None

**Equivalent ARM syntax and encoding**

```
SBCS Rd, Rd, Rm
```

| 31 | 30 | 29 | 28 | 27 | 26 | 25 | 24 | 23 | 22 | 21 | 20 | 19 | | 16 | 15 | | 12 | 11 | 10 | 9 | 8 | 7 | 6 | 5 | 4 | 3 | | 0 |
|----|----|----|----|----|----|----|----|----|----|----|----|----|---|----|----|---|----|----|----|---|---|---|---|---|---|---|---|---|
| 1 | 1 | 1 | 0 | 0 | 0 | 0 | 0 | 0 | 1 | 1 | 0 | 1 | Rd | | Rd | | | 0 | 0 | 0 | 0 | 0 | 0 | 0 | 0 | Rm | | |

**ARM Architecture Reference Manual**
ARM DUI 0100B

STMIA Rn!, <register_list>

**Description**   The STM (Store Multiple) instruction is useful as a block store instruction. Combined with LDM (load multiple), it allows efficient block copy.

The STMIA (Store Multiple Increment After) instruction stores a subset (or possibly all) of the general-purpose registers to sequential memory locations. The registers are stored in sequence:

- the lowest-numbered register first, to the lowest memory address (<start_address>)

- the highest-numbered register last, to the highest memory address (<end_address>)

The <start_address> is the value of the base register Rn.

Subsequent addresses are formed by incrementing the previous address by four. One address is produced for each register that is specified in <register_list>.

The <end_address> value is four less than the sum of the value of the base register and four times the number of registers specified in <register_list>.

Finally, the base register Rn is incremented by 4 times the numbers of registers in <register_list>.

| 15 | 14 | 13 | 12 | 11 | 10 | 8 | 7 | 0 |
|---|---|---|---|---|---|---|---|---|
| 1 | 1 | 0 | 0 | 0 | Rn | | register_list | |

### Operation

```
<start_address> = Rn
<end_address> = Rn + (Number_Of_Set_Bits_In(<register_list>) * 4) - 4
<address> = <start_address>
for i = 0 to 7
    if <register_list>[i] == 1
        Memory[<address>,4] = Ri
        <address> = <address> + 4
assert <end_address> == <address> - 4
Rn = Rn + (Number_Of_Set_Bits_In(<register_list>) * 4)
```

### Equivalent ARM syntax and encoding

STMIA Rn!, <register_list>

| 31 | 30 | 29 | 28 | 27 | 26 | 25 | 24 | 23 | 22 | 21 | 20 | 19 | 16 | 15 | 14 | 13 | 12 | 11 | 10 | 9 | 8 | 7 | 0 |
|---|---|---|---|---|---|---|---|---|---|---|---|---|---|---|---|---|---|---|---|---|---|---|---|
| 1 | 1 | 1 | 0 | 1 | 0 | 0 | 0 | 1 | 0 | 1 | 0 | Rn | | 0 | 0 | 0 | 0 | 0 | 0 | 0 | 0 | register_list | |

## ARM Architecture Reference Manual
ARM DUI 0100B

Architecture v4T only

| | |
|---|---|

**Exceptions**   Data Abort

**Qualifiers**   None

**Notes**   **Register Rn:** Specifies the base register used by <addressing_mode>.

**Operand restrictions:** If the base register Rn is specified in <register_list>, ~~and writeback is specified~~, the value of Rn stored for Rn is UNPREDICTABLE.

**Data Abort:** If a data abort is signalled, the value left in Rn is IMPLEMENTATION DEFINED, but is either the original base register value or the updated base register value.

**Non-word-aligned addresses:** Store multiple instructions ignore the least-significant two bits of <address> (the words are not rotated as for load word).

**Alignment:** If an implementation includes a System Control Coprocessor (see *Chapter 7, System Architecture and System Control Coprocessor*), and alignment checking is enabled, an address with bits[1:0] != 0b00 will cause an alignment exception.

**ARM Architecture Reference Manual**
ARM DUI 0100B

`STR Rd, [Rn, #5_bit_offset])`

**Description**    The STR (Store Register) instruction allows 32-bit data from a general-purpose register to be stored to memory. The addressing mode is useful for accessing structure (record) fields. With an offset of zero, the address produced is the unaltered value of the base register Rn.

STR stores a word from register Rd to memory. The memory address is calculated by adding 4 times the value of `<5_bit_offset>` to the value of register Rn. If the address is not word-aligned, the result is UNPREDICTABLE.

| 15 | 14 | 13 | 12 | 11 | 10 | 6 | 5 | 3 | 2 | 0 |
|----|----|----|----|----|----|---|---|---|---|---|
| 0 | 1 | 1 | 0 | 0 | 5_bit_offset | | Rn | | Rd | |

**Operation**
```
<address> = Rn + (5_bit_offset * 4)
<data> = Rd
if <address>[1:0] == 0b00
        Memory[<address>,4] = <data>
else
        Memory[<address>,4] = UNPREDICTABLE
```

**Exceptions**    Data Abort

**Qualifiers**    None

**Notes**    **Alignment:** If an implementation includes a System Control Coprocessor (see *Chapter 7, System Architecture and System Control Coprocessor*), and alignment checking is enabled, an address with bits[1:0] != 0b00 will cause an alignment exception.

### Equivalent ARM syntax and encoding

`STR Rd, [Rn, #5_bit_offset]`

| 31 | 30 | 29 | 28 | 27 | 26 | 25 | 24 | 23 | 22 | 21 | 20 | 19 | 16 | 15 | 12 | 11 | 10 | 9 | 8 | 7 | 6 | 2 | 1 | 0 |
|----|----|----|----|----|----|----|----|----|----|----|----|----|----|----|----|----|----|---|---|---|---|---|---|---|
| 1 | 1 | 1 | 0 | 0 | 1 | 0 | 1 | 1 | 0 | 0 | 0 | Rn | | Rd | | 0 | 0 | 0 | 0 | 0 | 5_bit_offset | | 0 | 0 |

**ARM Architecture Reference Manual**
ARM DUI 0100B

**Description**

This form of the STR (Store Register) instruction allows 32-bit data from a general-purpose register to be stored to memory. The addressing mode is useful for pointer + large offset arithmetic (use the MOV immediate to set the offset), and for accessing a single element of an array.

In this case, STR stores a word from register Rd to memory. The memory address is calculated by adding the value of register Rm to the value of register Rn. If the address is not word-aligned, the result is UNPREDICTABLE.

| 15 | 14 | 13 | 12 | 11 | 10 | 9 | 8 | | 6 | 5 | | 3 | 2 | | 0 |
|----|----|----|----|----|----|---|---|---|---|---|---|---|---|---|---|
| 0 | 1 | 0 | 1 | 0 | 0 | 0 | | Rm | | | Rn | | | Rd | |

**Operation**

```
<address> = Rn + Rm
<data> = Rd
if <address>[1:0] == 0b00
        Memory[<address>,4] = <data>
else
        Memory[<address>,4] = UNPREDICTABLE
```

**Exceptions**    Data Abort

**Qualifiers**    None

**Notes**

**Alignment:** If an implementation includes a System Control Coprocessor (see *Chapter 7, System Architecture and System Control Coprocessor*), and alignment checking is enabled, an address with bits[1:0] != 0b00 will cause an alignment exception.

---

**Equivalent ARM syntax and encoding**

STR Rd, [Rn, Rm]

| 31 | 30 | 29 | 28 | 27 | 26 | 25 | 24 | 23 | 22 | 21 | 20 | 19 | | 16 | 15 | | 12 | 11 | 10 | 9 | 8 | 7 | 6 | 5 | 4 | 3 | | 1 | 0 |
|----|----|----|----|----|----|----|----|----|----|----|----|----|---|----|----|---|----|----|----|---|---|---|---|---|---|---|---|---|---|
| 1 | 1 | 1 | 0 | 0 | 1 | 1 | 1 | 1 | 0 | 0 | 0 | | Rn | | | Rd | | 0 | 0 | 0 | 0 | 0 | 0 | 0 | 0 | | Rm | | |

---

**ARM Architecture Reference Manual**
ARM DUI 0100B

*Store word
SP-relative*

*Architecture v4T only*

STR Rd, [SP, #8_bit_offset]

**Description**  This form of the STR (Store Register) instruction allows 32-bit data from
a general-purpose register to be stored to memory. The addressing mode is useful
for accessing stack data.

In this case, STR stores a word from register Rd to memory. The memory address
is calculated by adding 4 times the value of <8_bit_offset> to the value of
the SP. If the address is not word-aligned, the result is UNPREDICTABLE.

| 15 | 14 | 13 | 12 | 11 | 10 | 9 | 8 | 7 | 6 | 0 |
|----|----|----|----|----|----|----|----|----|----|----|
| 1 | 0 | 0 | 1 | 0 | | Rd | | | 8_bit_offset | |

**Operation**

```
<address> = SP + (8_bit_offset * 4)
<data> = Rd
if <address>[1:0] == 0b00
        Memory[<address>,4] = Rd
else
        Memory[<address>,4] = UNPREDICTABLE
```

**Exceptions**  Data Abort

**Qualifiers**  None

**Notes**  **Alignment:** If an implementation includes a System Control Coprocessor
((see *Chapter 7, System Architecture and System Control Coprocessor*), and
alignment checking is enabled, an address with bits[1:0] != 0b00 will cause
an alignment exception.

---

**Equivalent ARM syntax and encoding**

STR Rd, [SP, #8_bit_offset]

| 31 | 30 | 29 | 28 | 27 | 26 | 25 | 24 | 23 | 22 | 21 | 20 | 19 | 18 | 17 | 16 | 15 | | 12 | 11 | 10 | 9 | | 2 | 1 | 0 |
|----|----|----|----|----|----|----|----|----|----|----|----|----|----|----|----|----|----|----|----|----|----|----|----|----|----|
| 1 | 1 | 1 | 0 | 0 | 1 | 0 | 1 | 1 | 0 | 0 | 0 | 1 | 1 | 0 | 1 | Rd | | | 0 | 0 | 8_bit_offset | | | 0 | 0 |

**ARM Architecture Reference Manual**
ARM DUI 0100B

**Description**

This form of the STRB (Store Register Byte) instruction allows 8-bit data from a general-purpose register to be stored to memory. The addressing mode is useful for accessing structure (record) fields. With an offset of zero, the address produced is the unaltered value of the base register Rn.

In this case, STRB stores a byte from the least-significant byte of register Rd to memory. The memory address is calculated by adding the value of <5_bit_offset> to the value of register Rn.

*Architecture v4T only*

| 15 | 14 | 13 | 12 | 11 | 10 | 6 | 5 | 3 | 2 | 0 |
|---|---|---|---|---|---|---|---|---|---|---|
| 0 | 1 | 1 | 1 | 0 | 5_bit_offset | | Rn | | Rd | |

**Operation**

```
<address> = Rn + 5_bit_offset
Memory[<address>,1] = Rd[7:0]
```

**Exceptions**    Data Abort

**Qualifiers**    None

**Equivalent ARM syntax and encoding**

STRB Rd, [Rn, #5_bit_offset]

| 31 | 30 | 29 | 28 | 27 | 26 | 25 | 24 | 23 | 22 | 21 | 20 | 19 | 16 | 15 | 12 | 11 | 10 | 9 | 8 | 7 | 6 | 2 | 1 | 0 |
|---|---|---|---|---|---|---|---|---|---|---|---|---|---|---|---|---|---|---|---|---|---|---|---|---|
| 1 | 1 | 1 | 0 | 0 | 1 | 0 | 1 | 1 | 1 | 0 | 0 | Rn | | Rd | | 0 | 0 | 0 | 0 | 0 | 5_bit_offset | | 0 | 0 |

**ARM Architecture Reference Manual**
ARM DUI 0100B

# STRB (2)

**Store byte
register offset**

*Architecture v4T only*

**Description**

This form of the STRB (Store Register Byte) instruction allows 8-bit data from a general-purpose register to be stored to memory. The addressing mode is useful for pointer + large offset arithmetic (use the MOV immediate to set the offset), and for accessing a single element of an array.

In this case, STRB stores a byte from the least-significant byte of register Rd to memory. The memory address is calculated by adding the value register Rm to the value of register Rn.

| 15 | 14 | 13 | 12 | 11 | 10 | 9 | 8 | | 6 | 5 | | 3 | 2 | | 0 |
|----|----|----|----|----|----|---|---|---|---|---|---|---|---|---|---|
| 0 | 1 | 0 | 1 | 0 | 1 | 0 | | Rm | | | Rn | | | Rd | |

**Operation**

```
<address> = Rn + Rm
Memory[<address>,1] = Rd[7:0]
```

**Exceptions**    Data Abort

**Qualifiers**    None

---

**Equivalent ARM syntax and encoding**

STRB Rd, [Rn, Rm]

| 31 | 30 | 29 | 28 | 27 | 26 | 25 | 24 | 23 | 22 | 21 | 20 | 19 | | 16 | 15 | | 12 | 11 | 10 | 9 | 8 | 7 | 6 | 5 | 4 | 3 | | | 0 |
|----|----|----|----|----|----|----|----|----|----|----|----|----|---|----|----|---|----|----|----|---|---|---|---|---|---|---|---|---|---|
| 1 | 1 | 1 | 0 | 0 | 1 | 1 | 1 | 1 | 1 | 0 | 0 | | Rn | | | Rd | | 0 | 0 | 0 | 0 | 0 | 0 | 0 | 0 | | | Rm | |

**ARM Architecture Reference Manual**
ARM DUI 0100B

**ARM**

# STRH (1) <sub>Thumb</sub>

Store halfword
immediate offset

Architecture v4T only

**Description**

This form of the STRH (Store Register Halfword) instruction allows 16-bit data from a general-purpose register to be stored to memory. The addressing mode is useful for accessing structure (record) fields. With an offset of zero, the address produced is the unaltered value of the base register Rn.

In this case, STRH stores a halfword from the least-significant halfword of register Rd to memory. The memory address is calculated by adding 2 times the value of <5_bit_offset> to the value of register Rn. If the address is not halfword-aligned, the result is UNPREDICTABLE.

| 15 | 14 | 13 | 12 | 11 | 10 | | | 6 | 5 | | 3 | 2 | | 0 |
|---|---|---|---|---|---|---|---|---|---|---|---|---|---|---|
| 1 | 0 | 0 | 0 | 0 | | 5_bit_offset | | | | Rn | | | Rd | |

**Operation**

```
<address> = Rn + (5_bit_offset * 2)
<data> = Rd
if <address>[1:0] == 0
      Memory[<address>,2] = <data>[15:0]
else
      Memory[<address>,2] = UNPREDICTABLE
```

**Exceptions**    Data Abort

**Qualifiers**    None

**Notes**    **Alignment:** If an implementation includes a System Control Coprocessor (see *Chapter 7, System Architecture and System Control Coprocessor*), and alignment checking is enabled, an address with bit[0] != 0 will cause an alignment exception.

**Equivalent ARM syntax and encoding**

STRH Rd, [Rn, #5_bit_offset]

| 31 | 30 | 29 | 28 | 27 | 26 | 25 | 24 | 23 | 22 | 21 | 20 | 19 | | | 16 | 15 | | | 12 | 11 | 10 | 9 | 8 | 7 | 6 | 5 | 4 | 3 | | | 0 |
|---|---|---|---|---|---|---|---|---|---|---|---|---|---|---|---|---|---|---|---|---|---|---|---|---|---|---|---|---|---|---|---|
| 1 | 1 | 1 | 0 | 0 | 0 | 0 | 0 | 1 | 1 | 1 | 0 | 0 | | Rn | | | Rd | | | 0 | 0 | 0 | O4 | 1 | 0 | 1 | 1 | | offset[3:0] | | |

STRH Rd, [Rn, Rm]

**Description**  This form of the STRH (Store Register Halfword) instruction allows 16-bit data from a general-purpose register to be stored to memory. The addressing mode is useful for pointer + large offset arithmetic (use the MOV immediate to set the offset), and accessing a single element of an array.

In this case, STRH stores a halfword from the least-significant halfword of register Rd to memory. The memory address is calculated by adding the value of register Rm to the value of register Rn. If the address is not halfword-aligned, the result is UNPREDICTABLE.

| 15 | 14 | 13 | 12 | 11 | 10 | 9 | 8 | | 6 | 5 | | 3 | 2 | | 0 |
|----|----|----|----|----|----|---|---|---|---|---|---|---|---|---|---|
| 0 | 1 | 0 | 1 | 0 | 0 | 1 | Rm | | | Rn | | | Rd | | |

**Operation**

```
<address> = Rn + Rm
<data> = Rd
if <address>[1:0] == 0
      Memory[<address>,2] = <data>[15:0]
else
      Memory[<address>,2] = UNPREDICTABLE
```

**Exceptions**  Data Abort

**Qualifiers**  None

**Notes**  **Alignment:** If an implementation includes a System Control Coprocessor (see *Chapter 7, System Architecture and System Control Coprocessor*), and alignment checking is enabled, an address with bit[0] != 0 will cause an alignment exception.

### Equivalent ARM syntax and encoding

STRH Rd, [Rn, Rm]

| 31 | 30 | 29 | 28 | 27 | 26 | 25 | 24 | 23 | 22 | 21 | 20 | 19 | | 16 | 15 | | 12 | 11 | | 8 | 7 | 6 | 5 | 4 | 3 | 2 | 1 | 0 |
|----|----|----|----|----|----|----|----|----|----|----|----|----|---|----|----|---|----|----|---|---|---|---|---|---|---|---|---|---|
| 1 | 1 | 1 | 0 | 0 | 0 | 0 | 1 | 1 | 0 | 0 | 0 | Rn | | | Rd | | | SBZ | | | 1 | 0 | 1 | 1 | Rm | | | |

**ARM Architecture Reference Manual**

ARM DUI 0100B

*Subtract immediate*

**Description**  This form of the SUB (Subtract) instruction subtracts a small constant value from the value of a register and stores the result in a second register.

In this case, SUB subtracts the value of the 3-bit immediate (values 0 to 7) from the value of register Rn, and stores the result in the destination register Rd. The condition code flags are updated (based on the result).

*Architecture v4T only*

| 15 | 14 | 13 | 12 | 11 | 10 | 9 | 8 | | 6 | 5 | | 3 | 2 | | 0 |
|----|----|----|----|----|----|---|---|---|---|---|---|---|---|---|---|
| 0 | 0 | 0 | 1 | 1 | 1 | 1 | 3_bit_immediate | | | Rn | | | Rd | | |

**Operation**
```
Rd = Rn - <3_bit_immed>
N Flag = Rd[31]
Z Flag = if Rd == 0 then 1 else 0
C Flag = NOT BorrowFrom(Rn - <3_bit_immed>)
V Flag = OverflowFrom(Rn - <3_bit_immed>)
```

**Exceptions**  None

**Qualifiers**  None

---

**Equivalent ARM syntax and encoding**

SUBS Rd, Rn, #<3_bit_immediate>

| 31 | 30 | 29 | 28 | 27 | 26 | 25 | 24 | 23 | 22 | 21 | 20 | 19 | | 16 | 15 | | 12 | 11 | 10 | 9 | 8 | 7 | 6 | 5 | 4 | 3 | 2 | | 0 |
|----|----|----|----|----|----|----|----|----|----|----|----|----|---|----|----|---|----|----|----|---|---|---|---|---|---|---|---|---|---|
| 1 | 1 | 1 | 0 | 0 | 0 | 1 | 0 | 0 | 1 | 0 | 1 | | Rn | | | Rd | | 0 | 0 | 0 | 0 | 0 | 0 | 0 | 0 | 0 | 0 | #3_bit_imm | |

**ARM Architecture Reference Manual**
ARM DUI 0100B

*Subtract
large immediate*

*Architecture v4T only*

SUB Rd, #<8_bit_immediate>

**Description**

This form of the SUB (Subtract) instruction subtracts a large constant value from the value of a register and stores the result back in the same register.

In this case, SUB subtracts the value of the 8-bit immediate (values 0 to 255) from the value of register Rd, and stores the result back in the register Rd.
The condition code flags are updated (based on the result).

| 15 | 14 | 13 | 12 | 11 | 10 | | 8 | 7 | | 0 |
|----|----|----|----|----|----|----|----|----|----|----|
| 0 | 0 | 1 | 1 | 1 | | Rd | | | 8_bit_immediate | |

**Operation**

```
Rd = Rd - <8_bit_immed>
N Flag = Rd[31]
Z Flag = if Rd == 0 then 1 else 0
C Flag = NOT BorrowFrom(Rd - <8_bit_immed>)
V Flag = OverflowFrom(Rd - <8_bit_immed>)
```

**Exceptions**     None

**Qualifiers**     None

**Equivalent ARM syntax and encoding**

SUBS Rd, Rd, #<8_bit_immediate>

| 31 | 30 | 29 | 28 | 27 | 26 | 25 | 24 | 23 | 22 | 21 | 20 | 19 | | 16 | 15 | | 12 | 11 | 10 | 9 | 8 | 7 | | 0 |
|----|----|----|----|----|----|----|----|----|----|----|----|----|----|----|----|----|----|----|----|----|----|----|----|----|
| 1 | 1 | 1 | 0 | 0 | 0 | 1 | 0 | 0 | 1 | 0 | 1 | | Rd | | | Rd | | 0 | 0 | 0 | 0 | | 8_bit_immediate | |

**ARM Architecture Reference Manual**
ARM DUI 0100B

This form of the SUB (Subtract) instruction subtracts the value of one register from the value of a second register and stores the result in a third register.

In this case, SUB subtracts the value of register Rm from the value of register Rn, and stores the result in the destination register Rd. The condition code flags are updated (based on the result).

| 15 | 14 | 13 | 12 | 11 | 10 | 9 | 8 | | 6 | 5 | | 3 | 2 | | 0 |
|---|---|---|---|---|---|---|---|---|---|---|---|---|---|---|---|
| 0 | 0 | 0 | 1 | 1 | 0 | 1 | | Rm | | | Rn | | | Rd | |

**Operation**

```
Rd = Rn - Rm
N Flag = Rd[31]
Z Flag = if Rd == 0 then 1 else 0
C Flag = NOT BorrowFrom(Rn - Rm)
V Flag = OverflowFrom(Rn - Rm)
```

**Exceptions**   None

**Qualifiers**   None

### Equivalent ARM syntax and encoding

`SUBS Rd, Rn, Rm`

| 31 | 30 | 29 | 28 | 27 | 26 | 25 | 24 | 23 | 22 | 21 | 20 | 19 | | 16 | 15 | | 12 | 11 | 10 | 9 | 8 | 7 | 6 | 5 | 4 | 3 | | 0 |
|---|---|---|---|---|---|---|---|---|---|---|---|---|---|---|---|---|---|---|---|---|---|---|---|---|---|---|---|---|
| 1 | 1 | 1 | 0 | 0 | 0 | 0 | 0 | 0 | 1 | 0 | 1 | | Rn | | | Rd | | 0 | 0 | 0 | 0 | 0 | 0 | 0 | 0 | | Rm | |

**ARM Architecture Reference Manual**
ARM DUI 0100B

# SUB (4)

Decrement
stack pointer

Architecture v4T only

SUB, SP, SP, #<7_bit_immediate>

**Description**

This form of the SUB (Subtract) instruction is used to increase the size of the stack.

In this case, SUB subtracts the value of the 7-bit immediate (values 0 to 127) and the value of the SP, and stores the result back in the SP.   *O to 1FC*

| 15 | 14 | 13 | 12 | 11 | 10 | 9 | 8 | 7 | 6 | 5 | | | 0 |
|----|----|----|----|----|----|----|----|----|----|----|----|----|----|
| 1 | 0 | 1 | 1 | 0 | 0 | 0 | 0 | 1 | | | 7_bit_immediate | | |

**Operation**     SP = SP - <7_bit_immed>  *<< 2*

**Exceptions**     None

**Qualifiers**     None

---

**Equivalent ARM syntax and encoding**

SUB SP, SP, #<7_bit_immediate>

| 31 | 30 | 29 | 28 | 27 | 26 | 25 | 24 | 23 | 22 | 21 | 20 | 19 | 18 | 17 | 16 | 15 | 14 | 13 | 12 | 11 | 10 | 9 | 8 | 7 | 6 | | 0 |
|----|----|----|----|----|----|----|----|----|----|----|----|----|----|----|----|----|----|----|----|----|----|----|----|----|----|----|----|
| 1 | 1 | 1 | 0 | 0 | 0 | 1 | 0 | 0 | 1 | 0 | 0 | 1 | 1 | 0 | 1 | 1 | 1 | 0 | 1 | 0 | 0 | 0 | 0 | 0 | | 7_bit_immediate | |

**ARM Architecture Reference Manual**
ARM DUI 0100B

**ARM**

| **Description** | The SWI instruction is used as an operating system service call. It can be used in two ways: | *Software Interrupt* |
|---|---|---|

1 Uses the 8-bit offset to indicate the OS service that is required.

2 Ignores the 8-bit field and indicates the service required with a general-purpose register.

*Architecture v4T only*

A SWI exception is generated, which is handled by an operating system to provide the requested service; see *2.5 Exceptions* on page 2-6.

| 15 | 14 | 13 | 12 | 11 | 10 | 9 | 8 | 7 | | 0 |
|---|---|---|---|---|---|---|---|---|---|---|
| 1 | 1 | 0 | 1 | 1 | 1 | 1 | 1 | | 8_bit_immediate | |

**Operation**

```
R14_svc = PC + ??
SPSR_svc = CPSR
CPSR[/]5= 0          ; begin ARM execution
CPSR[4:0] = 0b10011   ; enter Supervisor mode
CPSR[7] = 1           ; disable IRQ
PC = 0x08
```

**Exceptions**     None

**Qualifiers**     None

### Equivalent ARM syntax and encoding

SWI &lt;8_bit_immediate&gt;

| 31 | 30 | 29 | 28 | 27 | 26 | 25 | 24 | 23 | 22 | 21 | 20 | 19 | 18 | 17 | 16 | 15 | 14 | 13 | 12 | 11 | 10 | 9 | 8 | 7 | | 0 |
|---|---|---|---|---|---|---|---|---|---|---|---|---|---|---|---|---|---|---|---|---|---|---|---|---|---|---|
| 1 | 1 | 1 | 0 | 1 | 1 | 1 | 1 | 1 | 0 | 0 | 0 | 0 | 0 | 0 | 0 | 0 | 0 | 0 | 0 | 0 | 0 | 0 | 0 | | 8_bit_immediate | |

**ARM Architecture Reference Manual**
ARM DUI 0100B

# TST

**Test bits**

*Architecture v4T only*

TST Rn, Rm

**Description**

The TST (Test) instruction is used to determine if many bits of a register are all clear, or if at least one bit of a register is set.

TST performs a comparison by logically ANDing the value of register Rm from the value of register Rd. The condition code flags are updated (based on the result).

| 15 | 14 | 13 | 12 | 11 | 10 | 9 | 8 | 7 | 6 | 5 | | 3 | 2 | | 0 |
|----|----|----|----|----|----|---|---|---|---|---|---|---|---|---|---|
| 0 | 1 | 0 | 0 | 0 | 0 | 1 | 0 | 0 | 0 | | Rm | | | Rn | |

**Operation**

```
<alu_out> = Rn AND Rm
N Flag = <alu_out>[31]
Z Flag = if <alu_out> == 0 then 1 else 0
C Flag = UNAFFECTED
V Flag = UNAFFECTED
```

**Exceptions**     None

**Qualifiers**     None

**Equivalent ARM syntax and encoding**

TST Rn, Rm

| 31 | 30 | 29 | 28 | 27 | 26 | 25 | 24 | 23 | 22 | 21 | 20 | 19 | | 16 | 15 | | 12 | 11 | 10 | 9 | 8 | 7 | 6 | 5 | 4 | 3 | | | 0 |
|----|----|----|----|----|----|----|----|----|----|----|----|----|---|----|----|---|----|----|----|---|---|---|---|---|---|---|---|---|---|
| 1 | 1 | 1 | 0 | 0 | 0 | 0 | 1 | 0 | 0 | 0 | 1 | | Rn | | | SBZ | | 0 | 0 | 0 | 0 | 0 | 0 | 0 | 0 | | Rm | | |

**ARM Architecture Reference Manual**

ARM DUI 0100B

ARM

**7**

# System Architecture and System Control Coprocessor

# System Architecture and System Control Coprocessor

This chapter describes ARM system architecture, and the system control processor.

# System Architecture and System Control Coprocessor

## 7.1    Introduction

Implementations of the ARM architecture optionally incorporate:

- on-chip Memory Management Unit (MMU) (including Translation Lookaside Buffer(s) (TLB))
- Instruction and/or Data Cache (IDC)
- Write Buffer (WB)

If these functions are implemented, coprocessor 15 is used to control them. Coprocessor 15 is called the System Control Coprocessor or just CP15.

The MMU incorporates a two-level page table for virtual to physical address translation, and access permission attributes for each virtual to physical translation. The MMU page tables also contain cache and write buffer enables; therefore, if a cache or a write buffer is implemented, the MMU must also be implemented

## 7.2    CP15 Access

CP15 defines 16 registers. CP15 registers can only be accessed with MRC and MCR instructions (CDP, LDC and STC instructions to CP15 will cause an undefined instruction trap). The CRn field of MRC and MCR instructions specify the coprocessor register to access, and the CRm field and opcode_2 field are used to specify a particular action when addressing some registers.

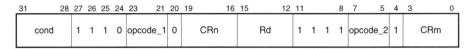

Opcode_1 should be zero (SBZ) for all CP15 instructions.

If a cache, MMU and Write Buffer are not implemented, CP15 will not be implemented, and all accesses to CP15 will cause undefined exceptions.

Any access to CP15 while the processor is in User mode will cause an undefined instruction exception.

An MRC instruction from coprocessor 15 to register 15 is UNPREDICTABLE.

## 7.3    CP15 Architectures

If a cache, MMU and Write Buffer are implemented, CP15 register 0 contains an architecture field that specifies a particular layout and functionality for the remaining registers.

Reading from CP15 register 0 returns an architecture and implementation-defined identification from the processor:

| 31          24 | 23          16 | 15          4 | 3          0 |
|----------------|----------------|---------------|--------------|
| Implementor    | Architecture version | Part number | Revision |

Bits[3:0]       contain the revision number for the processor

Bits[15:4]      contain a 3-digit part number in binary-coded decimal format

Bits[23:16]     contain the architecture version:

        0x00       Version 3 (see *7.5 ARMv3 System Control Coprocessor* on page 7-10)

        0x01       Version 4 (see *7.4 ARMv4 System Control Coprocessor* below)

Bits[31:24]     contain the ASCII code of an implementor's trademark:

        0x41 = A     ARM Ltd

        0x44 = D     Digital Equipment Corporation

## 7.4    ARMv4 System Control Coprocessor

ARM Architecture Version 4 System Control Coprocessor is designed to control a single combined instruction and data cache, or separate instructions and data caches, a write buffer, a prefetch buffer, and a virtual to physical address translator including combined instruction and data TLB or separate instruction and data TLB. CP15 also controls various system configuration signals.

| Register | Reads | Writes | Update Policy |
|----------|-------|--------|---------------|
| 0 | ID Register | UNPREDICTABLE | No update |
| 1 | Control | Control | Read Modify Write |
| 2 | Translation Table Base | Translation Table Base | |
| 3 | Domain Access Control | Domain Access Control | |
| 4 | UNPREDICTABLE | UNPREDICTABLE | |
| 5 | Fault Status | Fault Status | |
| 6 | Fault Address | Fault Address | |
| 7 | Cache Operations | Cache operations | Write only |
| 8 | TLB operations | TLB operations | Write only |
| 9 to 15 | UNPREDICTABLE | UNPREDICTABLE | |

*Table 7-1: ARMv4 CP15 register summary*

# System Architecture and System Control Coprocessor

## 7.4.1   Register 0: ID register

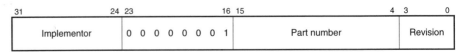

| 31 | 24 23 | 16 15 | 4 3 | 0 |
|---|---|---|---|---|
| Implementor | 0 0 0 0 0 0 0 1 | Part number | Revision | |

Reading from CP15 register 0 returns the implementation-defined identification for the processor. The CRm and opcode_2 fields are ignored when reading CP15 register 0, and SHOULD BE ZERO.

Bits[3:0]      contain the revision number for the processor

Bits[15:4]     contain a 3-digit part number in binary-coded decimal format
               (for example, 0x810 for ARM810)

Bits[23:16]    contain the architecture version
               (for example, 0x01 = Version 4)

Bits[31:24]    contain the ASCII code of an implementation trademark
               (0x41 = A = ARM Ltd.)

**Note**   Writing to CP15 register 0 is unpredictable.

## 7.4.2   Register 1: Control register

| 31 | 13 12 11 10 9 8 7 6 5 4 3 2 1 0 |
|---|---|
| DNM/SBZ | I Z F R S B L D P W C A M |

Reading from CP15 register 1 reads the control bits. The CRm and opcode_2 fields are IGNORED when reading CP15 register 1, and should be zero.

Writing to CP15 register 1 sets the control bits. The CRm and opcode_2 fields are not used when writing CP15 register 1, and should be zero.

All control bits are set to zero on reset. The control bits have the following functions:

M Bit 0      Memory Management Unit (MMU) Enable/Disable
             0 = MMU disabled
             1 = MMU enabled

A Bit 1      Alignment Fault Enable/Disable
             0 = Address alignment fault checking disabled
             1 = Address alignment fault checking enabled

C Bit 2      Instruction and data cache Enable/Disable
             If separate instruction and data caches are implemented, this bit
             controls only the data cache enable/disable, and the I bit (bit 12)
             controls the instruction cache enable/disable.
             0 = Instruction and data cache (IDC) disabled
             1 = Instruction and data cache (IDC) enabled

W Bit 3      Write buffer Enable/Disable
             0 = Write buffer disabled
             1 = Write buffer enabled

**ARM Architecture Reference Manual**

ARM  DDI 0100B

P Bit 4    32-bit/26-bit Exception handlers
Implementations that support 26-bit configurations (see *Chapter 5, The 26-bit Architectures*) use this bit to control the PROG32 signal.
0 = 26-bit exception handlers
1 = 32-bit exception handlers
This bit is UNPREDICTABLE on implementations that do not support 26-bit configurations, and should be 1.

D Bit 5    32-bit/26-bit data address range
Implementations that support 26-bit data spaces use this bit to control the DATA32 signal (see *Chapter 5, The 26-bit Architectures*).
0 = 26-bit data address checking enabled
1 = 26-bit data address checking disabled (32-bit data addresses)
This bit is UNPREDICTABLE on implementations that do not support 26-bit data spaces, and should be 1.

L Bit 6    IMPLEMENTATION DEFINED.

B Bit 7    Big endian/Little endian
0 = Little endian operation
1 = Big endian operation

S Bit 8    System protection
This bit modifies the MMU protection system.

R Bit 9    ROM protection
This bit modifies the MMU protection system.

F Bit 10   IMPLEMENTATION DEFINED

Z Bit 11   IMPLEMENTATION DEFINED

I Bit 12   Instruction cache enable/disable
0 = Instruction cache disabled
1 = Instruction cache enabled
If separate instruction and data caches are implemented, this bit controls the instruction cache enable/disable, and the C bit (bit 2) controls the data cache enable/disable. If a combined instruction and data cache is implemented, any writes to this bit are IGNORED, and reads return an UNPREDICTABLE value.

Bits 31:13    When read return an UNPREDICTABLE value, and when written SHOULD BE ZERO.

### Enabling the MMU

Care must be taken if the translated address differs from the untranslated address, as the instructions following the enabling of the MMU will have been fetched using no address translation and enabling the MMU may be considered as a branch with delayed execution. A similar situation occurs when the MMU is disabled. The correct code sequence for enabling and disabling the MMU is IMPLEMENTATION DEFINED.

If the cache and/or write buffer are enabled when the MMU is not enabled, the results are UNPREDICTABLE.

**ARM Architecture Reference Manual**

ARM DDI 0100B

### 7.4.3   Register 2: Translation table base register

| Translation Table Base | UNP/SBZP |
|---|---|

31                                                                                    14 13                                                    0

Reading from CP15 register 2 returns the pointer to the currently active first-level translation table in bits[31:14] and an unpredictable value in bits[13:0]. The CRm and opcode_2 fields are IGNORED when reading CP15 register 2, and SHOULD BE ZERO.

Writing to CP15 register 2 updates the pointer to the currently active first-level translation table from the value in bits[31:14] of the written value. Bits[13:0] must be written as zero. The CRm and opcode_2 fields are *ignored* when writing CP15 register 2, and SHOULD BE ZERO.

### 7.4.4   Register 3: Domain access control register

31 30 29 28 27 26 25 24 23 22 21 20 19 18 17 16 15 14 13 12 11 10  9  8  7  6  5  4  3  2  1  0

| D15 | D14 | D13 | D12 | D11 | D10 | D9 | D8 | D7 | D6 | D5 | D4 | D3 | D2 | D1 | D0 |
|---|---|---|---|---|---|---|---|---|---|---|---|---|---|---|---|

Reading from CP15 register 3 returns the value of the Domain Access Control Register. The CRm and opcode_2 fields are IGNORED when reading CP15 register 3, and SHOULD BE ZERO.

Writing to CP15 register 3 writes the value of the Domain Access Control Register. The CRm and opcode_2 fields are IGNORED when writing CP15 register 3, and SHOULD BE ZERO.

The Domain Access Control Register consists of sixteen 2-bit fields, each defining the access permissions for one of the 16 Domains (D15-D0). For the meaning of each field, see *7.9 Domains* on page 7-23.

### 7.4.5   Register 4: Reserved

Reading and writing CP15 register 4 is unpredictable.

**ARM Architecture Reference Manual**
ARM  DDI 0100B

## 7.4.6    Register 5: Fault status register

Reading CP15 register 5 returns the value of the Fault Status Register (FSR). The FSR contains the source of the last data fault. Note that only the bottom 9 bits are returned. The upper 23 bits are unpredictable. The FSR indicates the domain and type of access being attempted when an abort occurred.

| Bit 8 | returns zero |
|---|---|
| Bits 7:4 | specify which of the 16 domains (D15-D0) was being accessed when a fault occurred |
| Bits 3:0 | indicate the type of access being attempted. The encoding of these bits is shown in *Table 7-9: Priority encoding of fault status* on page 7-25. |

The FSR is only updated for data faults, not for prefetch faults. The CRm and opcode_2 fields are IGNORED when reading CP15 register 5, and SHOULD BE ZERO.

Writing CP15 register 5 sets the Fault Status Register to the value of the data written. This is useful for a debugger.to restore the value of the FSR. The upper 24 bits written should be zero (SBZ). The CRm and opcode_2 fields are IGNORED when writing CP15 register 5, and SHOULD BE ZERO.

## 7.4.7    Register 6: Fault address register

Reading CP15 register 6 returns the value of the Fault Address Register (FAR). The FAR holds the virtual address of the access which was attempted when a fault occurred. The FAR is only updated for data faults, not for prefetch faults. The CRm and opcode_2 fields are IGNORED when reading CP15 register 6, and SHOULD BE ZERO.

Writing CP15 register 6 sets the Fault Address Register to the value of the data written. This is useful for a debugger to restore the value of the FAR. The CRm and opcode_2 fields are IGNORED when writing CP15 register 6, and SHOULD BE ZERO.

# System Architecture and System Control Coprocessor

## 7.4.8 Register 7: Cache functions

Writing to CP15 register 7 is used to control caches and buffers.

An ARM implementation may include a combined instruction and data cache, or separate instruction and data caches. A write buffer, prefetch buffer and branch target cache may also be implemented and are controlled by this register. Several cache functions are defined, and the function to be performed is selected by the opcode_2 and CRm fields in the MCR instruction used to write CP15 register 7.

The Flush ID functions flush (invalidate) all cache data.

The Flush Entry functions may be implemented to flush more than a single entry, up to the entire cache.

The Clean D cache functions write out dirty data held in a writeback cache. They do not invalidate any cached data.

Any functions that are not relevant to a particular implementation are UNPREDICTABLE. All unused values of opcode_2 and CRm are UNPREDICTABLE.

Not all functions are provided by all implementations.

Reading CP15 register 7 is UNPREDICTABLE.

| Function | opcode_2 value | CRm value | Data | Instruction |
|---|---|---|---|---|
| Flush ID cache(s) | 0b000 | 0b0111 | SBZ | MCR p15, 0, Rd, c7, c7, 0 |
| Flush ID single entry | 0b001 | 0b0111 | IMP | MCR p15, 0, Rd, c7, c7, 1 |
| Flush I cache | 0b000 | 0b0101 | SBZ | MCR p15, 0, Rd, c7, c5, 0 |
| Flush I single entry | 0b001 | 0b0101 | IMP | MCR p15, 0, Rd, c7, c5, 1 |
| Flush D cache | 0b000 | 0b0110 | SBZ | MCR p15, 0, Rd, c7, c6, 0 |
| Flush D single entry | 0b001 | 0b0110 | IMP | MCR p15, 0, Rd, c7, c6, 1 |
| Clean ID cache | 0b000 | 0b1011 | SBZ | MCR p15, 0, Rd, c7, c11, 0 |
| Clean ID cache entry | 0b001 | 0b1011 | IMP | MCR p15, 0, Rd, c7, c11, 1 |
| Clean D cache | 0b000 | 0b1010 | SBZ | MCR p15, 0, Rd, c7, c10, 0 |
| Clean D cache entry | 0b001 | 0b1010 | IMP | MCR p15, 0, Rd, c7, c10, 1 |
| Clean and Flush ID cache | 0b000 | 0b1111 | SBZ | MCR p15, 0, Rd, c7, c15, 0 |
| Clean and Flush ID entry | 0b001 | 0b1111 | IMP | MCR p15, 0, Rd, c7, c15, 1 |
| Clean and Flush D cache | 0b000 | 0b1110 | SBZ | MCR p15, 0, Rd, c7, c14, 0 |
| Clean and Flush D entry | 0b001 | 0b1110 | IMP | MCR p15, 0, Rd, c7, c14, 1 |
| Flush Prefetch Buffer | 0b100 | 0b0101 | SBZ | MCR p15, 0, Rd, c7, c5, 4 |
| Drain Write Buffer | 0b100 | 0b1010 | SBZ | MCR p15, 0, Rd, c7, c10, 4 |
| Flush Branch Target Cache | 0b110 | 0b0101 | SBZ | MCR p15, 0, Rd, c7, c5, 6 |
| Flush Branch Target Entry | 0b111 | 0b0101 | IMP | MCR p15, 0, Rd, c7, c5, 7 |

*Table 7-2: Cache functions*

**ARM Architecture Reference Manual**
ARM DDI 0100B

## 7.4.9    Register 8: TLB functions

Writing to CP15 register 8 is used to control Translation Lookaside Buffers (TLBs). An ARM implementation may include a combined instruction and data TLB, or separate instruction and data TLBs. Several TLB functions are defined, and the function to be performed is selected by the opcode_2 and CRm fields in the MCR instruction used to write CP15 register 6.

Not all functions are provided by all implementations.

The Flush ID functions flush (invalidate) all TLB data.

The Flush I and Flush D functions are intended for use on implementations with split instruction and data TLBs; if used on an implementation with a combined TLB, the behaviour is as if a Flush ID function was used.

The Flush Entry functions may be implemented to flush from more than a single entry to the entire TLB.

Any functions that are not relevant to a particular implementation are UNPREDICTABLE. All unused values of opcode_2 and CRm are UNPREDICTABLE.

Reading CP15 register 8 is UNPREDICTABLE.

| Function | opcode_2 | CRm | Data | Instruction |
|---|---|---|---|---|
| Flush ID TLB(s) | 0b000 | 0b0111 | SBZ | MCR    p15,  0,  Rd,  c8,  c7,  0 |
| Flush ID single entry | 0b001 | 0b0111 | Virtual Address | MCR    p15,  0,  Rd,  c8,  c7,  1 |
| Flush I TLB | 0b000 | 0b0101 | SBZ | MCR    p15,  0,  Rd,  c8,  c5,  0 |
| Flush I single entry | 0b001 | 0b0101 | Virtual Address | MCR    p15,  0,  Rd,  c8,  c5,  1 |
| Flush D TLB | 0b000 | 0b0110 | SBZ | MCR    p15,  0,  Rd,  c8,  c6,  0 |
| Flush D single entry | 0b001 | 0b0110 | Virtual Address | MCR    p15,  0,  Rd,  c8,  c6,  1 |

*Table 7-3: TLB functions*

## 7.4.10   11-15: Reserved

Accessing (reading or writing) any of these registers is unpredictable.

# System Architecture and System Control Coprocessor

## 7.5 ARMv3 System Control Coprocessor

The ARM Architecture Version 3 System Control Coprocessor is designed to control a single combined instruction and data cache, a write buffer, and a virtual to physical address translator including combined instruction and data TLB.

CP15 also controls various system configuration signals:

| Register | Reads | Writes |
|----------|-------|--------|
| 0 | ID Register | UNPREDICTABLE |
| 1 | UNPREDICTABLE | Control |
| 2 | UNPREDICTABLE | Translation Table Base |
| 3 | UNPREDICTABLE | Domain Access Control |
| 4 | UNPREDICTABLE | UNPREDICTABLE |
| 5 | Fault Status | Flush TLB |
| 6 | Fault Address | Flush TLB Entry |
| 7 | UNPREDICTABLE | Flush Cache |
| 8 to 15 | UNDEFINED | UNDEFINED |

*Table 7-4: ARMv3 CP15 register summary*

### 7.5.1 Register 0: ID register

| 31 | 24 | 23 | 16 | 15 | 4 | 3 | 0 |
|----|----|----|----|----|----|----|----|
| Implementor | | 0 0 0 0 0 0 0 0 | | Part number | | Revision | |

Reading from CP15 register 0 returns an architecture and IMPLEMENTATION DEFINED identification for the processor. The CRm and opcode_2 fields are IGNORED when reading CP15 register 0, and SHOULD BE ZERO.

Bits[3:0]    contain the revision number for the processor

Bits[15:4]   contain a 3 digit part number in binary coded decimal format (for example, 0x700 for ARM700)

Bits[23:16]  contain the architecture version (for example, 0x00 = Version 3)

Bits[31:24]  contain the ASCII code of an implementor's trademark (for example, 0x41 = A = ARM Ltd.)

Writing to CP15 register 0 is unpredictable.

**ARM Architecture Reference Manual**

ARM  DDI 0100B

# System Architecture and System Control Coprocessor

## 7.5.2 Register 1: Control Register

| 31 | | 11 | 10 | 9 | 8 | 7 | 6 | 5 | 4 | 3 | 2 | 1 | 0 |
|---|---|---|---|---|---|---|---|---|---|---|---|---|---|
| UNP/SBZ | | | F | R | S | B | L | D | P | W | C | A | M |

Reading from CP15 register 1 is UNPREDICTABLE.

Writing to CP15 register 1 sets the control bits. The CRm and opcode_2 fields are IGNORED when writing CP15 register 1, and SHOULD BE ZERO.

All control bits are set to zero on reset. The control bits have the following functions:

M Bit 0
: Memory Management Unit (MMU) Enable/Disable
  0 = MMU disabled
  1 = MMU enabled

A Bit 1
: Alignment Fault Enable/Disable
  0 = Address alignment fault checking disabled
  1 = Address alignment fault checking enabled

C Bit 2
: Instruction and data cache Enable/Disable
  0 = Instruction and data cache (IDC) disabled
  1 = Instruction and data cache (IDC) enabled

W Bit 3
: Write buffer Enable/Disable
  0 = Write buffer disabled
  1 = Write buffer enabled

P Bit 4
: 32-bit/26-bit Exception handlers
  Implementations that support 26-bit configurations use this bit to control the PROG32 signal (see *Chapter 5, The 26-bit Architectures*)
  0 = 26-bit exception handlers
  1 = 32-bit exception handlers

D Bit 5
: 32-bit/26-bit data address range
  Implementations that support 26-bit data spaces use this bit to control the DATA32 signal (*Chapter 5, The 26-bit Architectures*)
  0 = 26-bit data address checking enabled
  1 = 26-bit data address checking disabled (32-bit data addresses)

L Bit 6
: IMPLEMENTATION DEFINED.

B Bit 7
: Big endian/Little endian
  0 = Little endian operation
  1 = Big endian operation

S Bit 8
: System protection
  This bit modifies the MMU protection system.

R Bit 9
: ROM protection
  This bit modifies the MMU protection system.

F Bit 10
: IMPLEMENTATION DEFINED

Bits 31:11
: When read, these bits return an UNPREDICTABLE value; when written, SHOULD BE ZERO.

# System Architecture and System Control Coprocessor

**Enabling the MMU**

Care must be taken if the translated address differs from the untranslated address, as the instructions following the enabling of the MMU will have been fetched using no address translation, and enabling the MMU may be considered as a branch with delayed execution. A similar situation occurs when the MMU is disabled.

The correct code sequence for enabling and disabling the MMU is IMPLEMENTATION DEFINED.

If the cache and/or write buffer are enabled when the MMU is not enabled, the results are UNPREDICTABLE.

## 7.5.3 Register 2: Translation Table Base Register

| 31 | 14 13 | 0 |
|---|---|---|
| Translation Table Base | | UNP/SBZ |

Reading from CP15 register 2 is UNPREDICTABLE.

Writing to CP15 register 2 updates the pointer to the currently active first-level translation table from the value in bits[31:14] of the written value. Bits[13:0] must be written as zero. The CRm and opcode_2 fields are IGNORED when writing CP15 register 2, and SHOULD BE ZERO.

## 7.5.4 Register 3: Domain Access Control Register

| 31 30 | 29 28 | 27 26 | 25 24 | 23 22 | 21 20 | 19 18 | 17 16 | 15 14 | 13 12 | 11 10 | 9 8 | 7 6 | 5 4 | 3 2 | 1 0 |
|---|---|---|---|---|---|---|---|---|---|---|---|---|---|---|---|
| D15 | D14 | D13 | D12 | D11 | D10 | D9 | D8 | D7 | D6 | D5 | D4 | D3 | D2 | D1 | D0 |

Reading from CP15 register 3 is UNPREDICTABLE.

Writing to CP15 register 3 writes the value of the Domain Access Control Register. The CRm and opcode_2 fields are IGNORED when writing CP15 register 3, and SHOULD BE ZERO.

The Domain Access Control Register consists of sixteen 2-bit fields, each of which defines the access permissions for one of the sixteen Domains (D15-D0).
For the meaning of each field, see *7.9 Domains* on page 7-23.

## 7.5.5 Register 4: Reserved

Reading and writing CP15 register 4 is UNPREDICTABLE.

**ARM Architecture Reference Manual**
ARM DDI 0100B

## 7.5.6 Register 5: Fault Status Register and Flush TLB

| 31 | | | 9 8 7 | 4 3 | 0 |
|---|---|---|---|---|---|
| UNP/SBZ | | | 0 | Domain | Status |

Reading CP15 register 5 returns the value of the Fault Status Register (FSR). The FSR contains the source of the last data fault. Note that only the bottom 9 bits are returned. The upper 23 bits are UNPREDICTABLE. The FSR indicates the domain and type of access being attempted when an abort occurred:

Bit 8      returns zero

Bits 7:4      specify which of the 16 domains (D15-D0) was being accessed when a fault occurred

Bits 3:0      indicate the type of access being attempted

The encoding is shown in *Table 7-9: Priority encoding of fault status* on page 7-25. The FSR is only updated for data faults, not for prefetch faults. The CRm and opcode_2 fields are IGNORED when reading CP15 register 5, and SHOULD BE ZERO.

Writing CP15 register 5 flushes the TLB. An ARMv3 implementation may only include a combined instruction and data TLB, and not separate instruction and data TLBs. The data written to the register is IGNORED, and SHOULD BE ZERO.

## 7.5.7 Register 6: Fault Address Register and Flush TLB Entry

| 31 | 0 |
|---|---|
| Fault address | |

Reading CP15 register 6 returns the value of the Fault Address Register (FAR). The FAR holds the virtual address of the access which was attempted when a fault occurred. The FAR is only updated for data faults, not for prefetch faults. The CRm and opcode_2 fields are IGNORED when reading CP15 register 6, and SHOULD BE ZERO.

Writing CP15 register 6 flushes a single entry from the TLB. An ARMv3 implementation may only include a combined instruction and data TLB, and not separate instruction and data TLBs. The data written to the register is the virtual address to be flushed.

## 7.5.8 Register 7: Flush Cache

Reading CP15 register 7 is UNPREDICTABLE.
Writing to CP15 register 7 is used to flush the instruction and data cache. An ARMv3 implementation may only include a combined instruction and data cache, and not separate instruction and data caches. The data written to the register is IGNORED, and SHOULD BE ZERO.

## 7.5.9 Registers 8-15: Reserved

Accessing (reading or writing) any of these registers will cause an undefined instruction exception.

# System Architecture and System Control Coprocessor

## 7.6 Memory Management Unit (MMU) Architecture

### 7.6.1 Overview

The ARM MMU performs two primary functions:

- it translates virtual addresses into physical addresses
- it controls memory access permissions

The MMU hardware required to perform these functions consists of:

- at least one Translation Lookaside Buffer (TLB)
- access control logic
- translation-table-walking logic

For implementations with separate Instruction and Data caches, separate TLBs for instruction and data are also likely.

**The translation lookaside buffer**

The TLB caches virtual to physical address translations and access permissions for each translation. If the TLB contains a translated entry for the virtual address, the access control logic determines whether access is permitted. If access is permitted, the MMU outputs the appropriate physical address corresponding to the virtual address. If access is not permitted, the MMU signals the CPU to abort.

If the TLB misses (it does not contain a translated entry for the virtual address), the translation table walk hardware is invoked to retrieve the translation and access permission information from a translation table in physical memory. Once retrieved, the information is placed into the TLB, possibly overwriting an existing entry.

**Memory accesses**

The MMU supports memory accesses based on *sections* or *pages*:

Sections       are comprised of 1MB blocks of memory

Pages          Two different page sizes are supported:

        *small pages*          consist of 4kB blocks of memory

        *large pages*          consist of 64kB blocks of memory

Sections and large pages are supported to allow mapping of a large region of memory while using only a single entry in the TLB. Additional access control mechanisms are extended within small pages to 1kB *sub-pages* and within large pages to 16kB *sub-pages*.

**Translation table**

The translation table held in main memory has two levels:

first-level table          holds both section translations and pointers to *second-level tables*

second-level tables        hold both large and small page translations

---

**ARM Architecture Reference Manual**
ARM DDI 0100B

**Domains**

The MMU also supports the concept of *domains.* These are areas of memory that can be defined to possess individual access rights. The Domain Access Control Register is used to specify access rights for up to 16 separate domains.

When the MMU is turned off (as happens on reset), the virtual address is output directly as the physical address, and no memory access permission checks are performed.

It is UNPREDICTABLE if two TLB entries address overlapping areas of memory. This can occur if the TLB is not flushed after memory is re-mapped with different-sized pages (leaving an old mapping with different sizes in the TLB, and a new mapping gets loaded into a different TLB location).

## 7.6.2 Translation process

The MMU translates virtual addresses generated by the CPU into physical addresses to access external memory, and also derives and checks the access permission. There are three routes by which the address translation (and hence permission check) takes place. The route taken depends on whether the address in question has been marked as a section-mapped access or a page-mapped access; and there are two sizes of page-mapped access (large pages and small pages).

However, the translation process always starts out in the same way, as described below, with a first-level fetch. A section-mapped access only requires a first level fetch, but a page-mapped access also requires a *second-level* fetch.

## 7.6.3 Translation table base

The translation process is initiated when the on-chip TLB does not contain an entry for the requested virtual address. The Translation Table Base Register points to the base of the first-level table. Only bits 31 to 14 of the Translation Table Base Register are significant; bits 13 to 0 should be zero. Therefore, the first-level page table must reside on a 16Kbyte boundary.

## 7.6.4 First-level fetch

Bits 31:14 of the Translation Table Base register are concatenated with bits 31:20 of the virtual address to produce a 30-bit address as illustrated in *Figure 7-1: Accessing the translation table first-level descriptors* on page 7-16. This address selects a four-byte translation table entry which is a first-level descriptor for a section or a pointer to a second-level page table.

# System Architecture and System Control Coprocessor

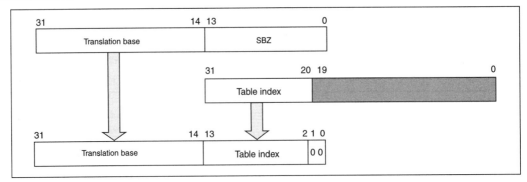

*Figure 7-1: Accessing the translation table first-level descriptors*

## 7.6.5 First-level descriptors

The first-level descriptor may define either a section descriptor or a pointer to a second-level page table and its format varies accordingly. *Figure 7-2: First-level descriptor format* shows the format, bits[1:0] indicate the descriptor type and validity.

Accessing a descriptor that has bits[1:0] = 0b00 generates a translation fault (see *7.10 Aborts* on page 7-24).

Accessing a descriptor that has bits[1:0] = 0b11 is UNPREDICTABLE.

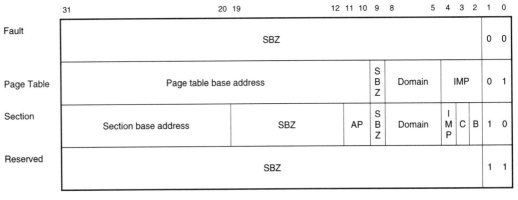

*Figure 7-2: First-level descriptor format*

## 7.6.6 Section descriptor and translating section references

If the first-level descriptor is a section descriptor, the fields have the following meanings:

| | |
|---|---|
| Bits 1:0 | Identify the type of descriptor (0b10 marks a section descriptor). |
| Bit 3:2 | The cachable and bufferable bits. See *7.7 Cache and Write Buffer Control* on page 7-22. |
| Bit 4 | The meaning of this bit is IMPLEMENTATION DEPENDENt. |
| Bits 8:5 | The domain field specifies one of the sixteen possible domains for all the pages controlled by this descriptor. |
| Bits 9 | This bit is not currently used, and SHOULD BE ZERO. |
| Bits 11:10 | Access permissions. These bits control the access to the section. See *Table 7-7: Access permissions* on page 7-23 for the interpretation of these bits. |
| Bits 19:12 | These bits are not currently used, and SHOULD BE ZERO. |
| Bits 31:20 | The Section Base Address forms the top 12 bits of the physical address. |

*Figure 7-3: Section translation* illustrates the complete section translation sequence. Note that the access permissions contained in the first-level descriptor must be checked before the physical address is generated. The sequence for checking access permissions is described in *7.8 Access Permissions* on page 7-22.

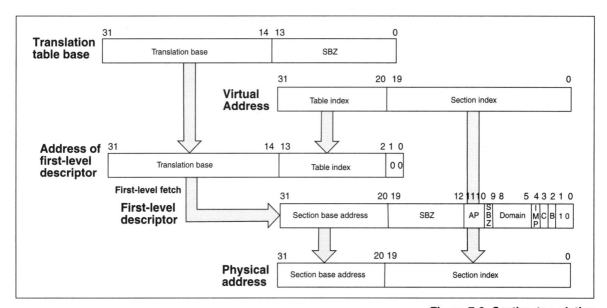

***Figure 7-3: Section translation***

# System Architecture and System Control Coprocessor

## 7.6.7 Page table descriptor

If the first-level descriptor is a page table descriptor, the fields have the following meanings:

Bits 1:0      Identify the type of descriptor (0b01 marks a page table descriptor).

Bit 4:2      The meaning of these bits is IMPLEMENTATION DEPENDENT.

Bits 8:5      The domain field specifies one of the sixteen possible domains for all the pages controlled by this descriptor.

Bits 9      This bit is not currently used, and SHOULD BE ZERO.

Bits 31:10      The Page Table Base Address is a pointer to a second-level page table, giving the base address for a second level fetch to be performed. Second level page tables must be aligned on a 1Kbyte boundary.

If a page table descriptor is returned from the first-level fetch, a second-level fetch is initiated to retrieve a second level descriptor, as shown in *Figure 7-4: Accessing the translation table second-level descriptors*.

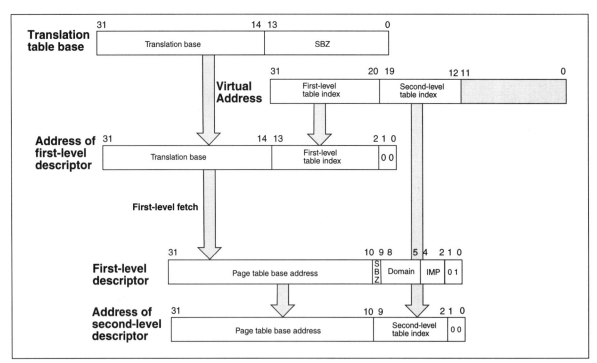

*Figure 7-4: Accessing the translation table second-level descriptors*

**ARM Architecture Reference Manual**

ARM DDI 0100B

## 7.6.8  Second-level descriptor

The second-level descriptor may define either a large page or a small page access. *Table 7-5: Second-level descriptor format* shows the format; bits[1:0] indicate the descriptor type and validity.

Accessing a descriptor that has bits[1:0] = 0b00 generates a translation fault (see *7.10 Aborts* on page 7-24).

Accessing a descriptor that has bits[1:0] = 0b11 is UNPREDICTABLE.

| | 31 ... 16 | 15 ... 12 | 11 10 | 9 8 | 7 6 | 5 4 | 3 | 2 | 1 | 0 |
|---|---|---|---|---|---|---|---|---|---|---|
| Fault | SBZ | | | | | | | | 0 | 0 |
| Large page | Large page base address | SBZ | AP3 | AP2 | AP1 | AP0 | C | B | 0 | 1 |
| Small page | Small page base address | | AP3 | AP2 | AP1 | AP0 | C | B | 1 | 0 |
| Reserved | SBZ | | | | | | | | 1 | 1 |

***Table 7-5: Second-level descriptor format***

The fields in both large and small pages have the following meanings:

Bits 1:0    identifies the type of descriptor

Bits 2:3    The cachable and bufferable bits. See *7.7 Cache and Write Buffer Control* on page 7-22.

Bits 11:4   Access permissions
These bits control access to the page. See *Table 7-7: Access permissions* on page 7-23 for the interpretation of these bits.

Both large and small pages are split into four sub-pages:

AP0         encodes the access permissions for the first sub-page

AP1         encodes the access permissions for the second sub-page

AP2         encodes the access permissions for the third sub-page

AP3         encodes the access permissions for the fourth (last) sub-page

Bits 15:12  are not currently used for large pages, and must be zero

Bits 31:12  are used to form the corresponding bits of the physical address (small pages)

Bits 31:16  are used to form the corresponding bits of the physical address (large pages)

# System Architecture and System Control Coprocessor

## 7.6.9    Translating large page references

*Figure 7-5: Large page translation* shows the complete translation sequence for a 64Kbyte large page.

**Note**    As the upper four bits of the Page Index and low-order four bits of the Second-level Table Index overlap, each page table entry for a large page must be duplicated 16 times (in consecutive memory locations) in the page table.

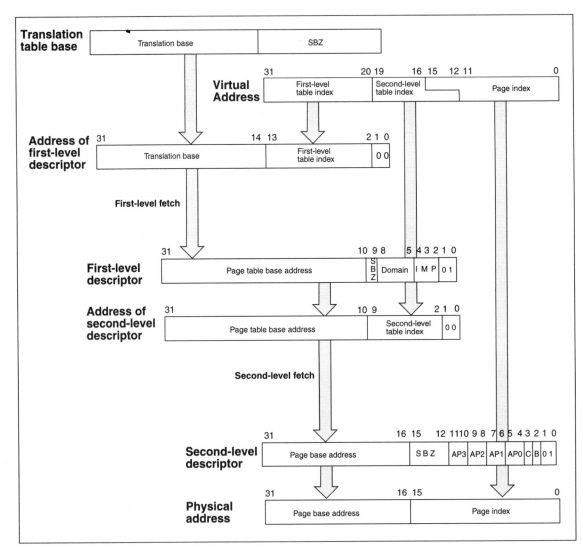

**Figure 7-5: Large page translation**

**ARM Architecture Reference Manual**

ARM  DDI 0100B

## 7.6.10 Translating small page references

*Figure 7-6: Small page translation* shows the complete translation sequence for a 4Kbyte small page.

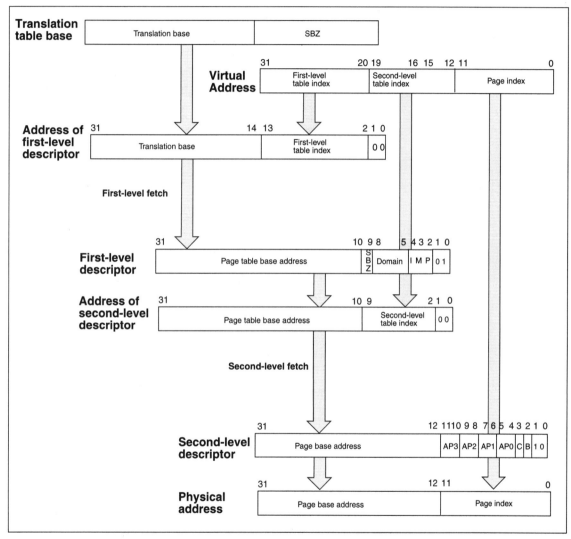

*Figure 7-6: Small page translation*

# System Architecture and System Control Coprocessor

## 7.7    Cache and Write Buffer Control

The ARM memory system is controlled by two attributes which are individually selectable for each virtual page:

Cacheable   This attribute indicates that data in the page may be cached, so that subsequent accesses may not access main memory. Cacheable also indicates that instruction speculative prefetching beyond the current point of execution may be performed. The cache implementation may use a write-back or a write-through policy (or a choice of either for individual virtual pages).

Bufferable   This attribute indicates that data in the page may be stored in the write buffer, allowing faster write operations for processors that operate faster than main memory. The write buffer may not preserve strict write ordering, and may not ensure that multiple writes to the same location result in multiple off-chip writes.

The Cacheable and Bufferable bits in the Section and Page descriptors control caching and buffering.

| C | B | Meaning |
|---|---|---------|
| 0 | 0 | Uncached, Unbuffered |
| 0 | 1 | Uncached, Buffered |
| 1 | 0 | Cached, Unbuffered or Writethrough cached, Buffered |
| 1 | 1 | Cached, Buffered or Writeback cached, Buffered |

*Table 7-6: Cache and bufferable bit meanings*

Implementations that offer both writethrough or writeback caching use the 0b10 value to specify writethrough caching, and the 0b11 value to specify writeback caching. Implementations that only offer one type of cache behaviour (writeback or writethrough) use the C and B bits strictly as cache enable and write buffer enable respectively.

Note that writeback cache implementations that do not also support writethrough caching, may not provide cached, unbuffered memory (as the writeback cache effectively buffers writes).

## 7.8    Access Permissions

The access permission bits in section and page descriptors control access to the corresponding section or page. The access permissions are modified by the System (S) and ROM (R) control bits. *Table 7-7: Access permissions* on page 7-23 describes the meaning of the access permission bits in conjunction with the S and R bits. If an access if made to an area of memory without the required permission, a Permission Fault is raised; see *7.10 Aborts* on page 7-24.

**ARM Architecture Reference Manual**

ARM DDI 0100B

| AP | S | R | Permissions | |
|----|---|---|-------------|--|
| | | | **Supervisor** | **User** |
| 00 | 0 | 0 | No Access | No Access |
| 00 | 1 | 0 | Read Only | No Access |
| 00 | 0 | 1 | Read Only | Read Only |
| 00 | 1 | 1 | UNPREDICTABLE | |
| 01 | x | x | Read/Write | No Access |
| 10 | x | x | Read/Write | Read Only |
| 11 | x | x | Read/Write | Read/Write |

*Table 7-7: Access permissions*

## 7.9 Domains

A domain is a collection of sections, large pages and small pages. The ARM architecture supports 16 domains; access to each domain is controlled by a 2-bit field in the Domain Access Control Register. Each field allows the access to an entire domain to be enabled and disabled very quickly, so that whole memory areas can be swapped in and out of virtual memory very efficiently.

Two kinds of domain access are supported:

Clients — are users of domains (execute programs, access data), and are guarded by the access permissions of the individual sections and pages that make up the domain

Managers — control the behaviour of the domain (the current sections and pages in the domain, and the domain access), and are not guarded by the access permissions of individual sections and pages in the domain

One program can be a client of some domains, and a manager of some other domains, and have no access to the remaining domains. This allows very flexible memory protection for programs that access different memory resources. *Table 7-8: Domain Access Values* illustrates the encoding of the bits in the Domain Access Control Register.

| Value | Access Types | Description |
|-------|--------------|-------------|
| 0b00 | No Access | Any access will generate a domain fault |
| 0b01 | Client | Accesses are checked against the access permission bits in the section or page descriptor |
| 0b10 | Reserved | Using this value has unpredictable results |
| 0b11 | Manager | Accesses are not checked against the access permission bits in the section or page descriptor, so a permission fault cannot be generated |

*Table 7-8: Domain Access Values*

# System Architecture and System Control Coprocessor

## 7.10 Aborts

The mechanisms that can cause the ARM processor to halt execution because of memory access restrictions are:

MMU fault   The MMU detects the restriction and signals the processor.

External abort   The external memory system signals an illegal memory access.

Collectively, MMU faults and external aborts are just called aborts. Accesses that cause aborts are said to be aborted.

If the memory request that is aborted is an instruction fetch, then a Prefetch Abort Exception is raised if and when the processor attempts to execute the instruction corresponding to the illegal access. If the aborted access is a data access, a Data Abort Exception is raised. See *2.5 Exceptions* on page 2-6 for more information about Prefetch and Data Aborts.

## 7.11 MMU Faults

The MMU generates four types of faults:

- Alignment Fault
- Translation Fault
- Domain Fault
- Permission Fault

The memory system may abort three types of access:

- Line Fetches
- Memory Accesses (uncached or unbuffered accesses)
- Translation Table Accesses

Aborts that are detected by the MMU are stopped before any external memory access takes place. It is the responsibility of the external system to stop external accesses that cause external aborts.

The System Control coprocessor contains two registers which are updated when a data access is aborted. These registers are not updated for prefetch aborts, as the aborted instruction may not be executed due to changes in program flow.

**ARM Architecture Reference Manual**

ARM DDI 0100B

## 7.11.1   Fault Address Register (FAR) and Fault Status Register (FSR)

Aborts resulting from data accesses (data aborts) are immediately acted upon by the CPU. The Fault Status Register (FSR) is updated with a 4-bit Fault Status (FS[3:0]) and the domain number of the access. In addition, the virtual address which caused the data abort is written into the Fault Address Register (FAR). If a data access simultaneously generates more than one type of data abort, they are prioritised in the order given in *Table 7-9: Priority encoding of fault status*.

Aborts arising from instruction fetches are simply flagged as the instruction enters the instruction pipeline. Only when (and if) the instruction is executed does it cause a prefetch abort; a prefetch abort is not acted upon if the instruction is not used (e.g. it is branched around). Because instruction prefetch aborts may or may not be acted upon, the FSR and FAR are not updated (the value of the PC saved in R14_abt after the exception occurs can be used to calculate the fault address).

| Priority | Sources | | FS[3:0] | Domain[3:0] | FAR |
|---|---|---|---|---|---|
| Highest | Terminal Exception | | 0b0010 | invalid | IMPLEMENTATION DEFINED |
| | Vector Exception | | 0b0000 | invalid | valid |
| | Alignment | | 0b00x1 | invalid | valid |
| | External Abort on Translation | First level | 0b1100 | invalid | valid |
| | | Second level | 0b1110 | valid | valid |
| | Translation | Section | 0b0101 | invalid | valid |
| | | Page | 0b0111 | valid | valid |
| | Domain | Section | 0b1001 | valid | valid |
| | | Page | 0b1011 | valid | valid |
| | Permission | Section | 0b1101 | valid | valid |
| | | Page | 0b1111 | valid | valid |
| | External Abort on Linefetch | Section | 0b0100 | valid | valid |
| | | Page | 0b0110 | valid | valid |
| Lowest | External Abort on Non-linefetch | Section | 0b1000 | valid | valid |
| | | Page | 0b1010 | valid | valid |

*Table 7-9: Priority encoding of fault status*

**Notes**

1   Alignment faults may write either 0b0001 or 0b0011 into FS[3:0].

2   Invalid values in Domain[3:0] occur because the fault is raised before a valid domain field has been loaded.

3   Any abort masked by the priority encoding may be regenerated by fixing the primary abort and restarting the instruction.

4   The FS[3:0] encoding for Vector Exception breaks from the pattern that FS[0] is zero for all external aborts.

# System Architecture and System Control Coprocessor

### 7.11.2 Fault-checking sequence

The sequence by which the MMU checks for access faults is slightly different for Sections and Pages. *Figure 7-7: Sequence for checking faults* on page 7-27 illustrates the sequence for both types of access. The sections and figures that follow describe the conditions that generate each of the faults.

### 7.11.3 Vector Exceptions

When the processor is in a 32-bit configuration (PROG32 is active) and in a 26-bit mode (CPSR[4] == 0), data access (but not instruction fetches) to the hard vectors (address 0x0 to 0x1f) will cause a data abort, known as a vector exception. See *7.11.3 Vector Exceptions* on page 7-26 for a full description. It is IMPLEMENTATION DEFINED if vector exceptions are generated when the MMU is not enabled.

### 7.11.4 Alignment fault

If Alignment Faults are enabled, an alignment fault will be generated on any data word access whose address is not word-aligned (virtual address bits [1:0] != 0b00), or any halfword access that is not halfword-aligned (virtual address bit[0] != 0). Alignment faults will not be generated on any instruction fetch, or on any byte access.

Note that if the access generates an alignment fault, the access will be aborted without reference to further permission checks. It is IMPLEMENTATION DEFINED if alignment exceptions are generated when the MMU is not enabled.

### 7.11.5 Translation fault

There are two types of translation fault:

Section     is generated if the first-level descriptor is marked as invalid. This happens if bits[1:0] of the descriptor are both 0.

Page        is generated if the second-level descriptor is marked as invalid. This happens if bits[1:0] of the descriptor are both 0.

### 7.11.6 Domain fault

There are two types of domain fault:

- Section
- Page

In both cases, the first descriptor holds the 4-bit Domain field which selects one of the sixteen 2-bit domains in the Domain Access Control Register. The two bits of the specified domain are then checked for access permissions as detailed in *Table 7-8: Domain Access Values* on page 7-23.

In the case of a section, the domain is checked when the first-level descriptor is returned, and in the case of a page, the domain is checked when the second-level descriptor is returned. If the specified access is marked as No Access in the Domain Access Control Register, either a Section Domain Fault or Page Domain Fault occurs.

**ARM Architecture Reference Manual**
ARM  DDI 0100B

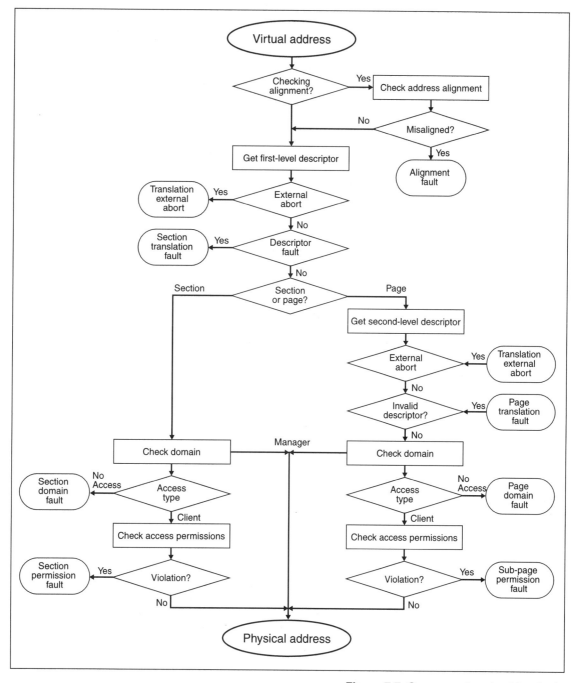

*Figure 7-7: Sequence for checking faults*

### 7.11.7 Permission fault

There are section permission faults and sub-page permission faults.

Permission faults are checked at the same time as domain faults. If the 2-bit domain field returns client (01), the permission access check is invoked as follows:

Section
If the first-level descriptor defines a section access, the AP bits of the descriptor define whether or not the access is allowed, according to *Table 7-7: Access permissions* on page 7-23. If the access is not allowed, a Section Permission fault is generated.

Sub-page
If the first-level descriptor defines a page-mapped access, the second-level descriptor specifies four access permission fields (ap3, ap2, ap1, ap0) each corresponding to one quarter of the page. For small pages, ap3 is selected by the top 1kB of the page, and ap0 is selected by the bottom 1kB of the page. For large pages, ap3 is selected by the top 16kB of the page, and ap0 is selected by the bottom 16kB of the page. The selected AP bits are then interpreted in exactly the same way as for a section, (see *Table 7-7: Access permissions* on page 7-23) the only difference being that the fault generated is a Sub-page Permission fault.

## 7.12 External Aborts

In addition to the MMU faults, the ARM Architecture defines an external abort pin which may be used to flag an error on an external memory access. However, not all accesses can be aborted in this way, so this pin must be used with great care. The following accesses may be externally aborted and restarted safely:

- Reads
- Unbuffered writes
- First-level descriptor fetch
- Second-level descriptor fetch
- Multi-bus master semaphores

A linefetch may be safely aborted on any word in the line transfer. If the abort happens on data that has been requested by the processor (rather than data that is being fetched as the remainder of a cache line), the access will be aborted. Any data transferred that is not immediately accessed (the remainder of the cache line) will only be aborted when it is accessed.

It is IMPLEMENTATION-DEFINED if the FAR points to the start address of the cache line, or the address that generated the abort.

Buffered writes cannot be externally aborted. Therefore, the system must be configured such that it does not do buffered writes to areas of memory which are capable of flagging an external abort, or a different mechanism should be used to signal the abort (an interrupt for example).

The contents of a memory location that causes an abort is UNPREDICTABLE after the abort.

**ARM Architecture Reference Manual**
ARM DDI 0100B

## 7.13 System-level Issues

This section lists a number of issues that need to be addressed by the system designer and operating systems to provide an ARMv4 compatible system.

### 7.13.1 Memory systems, write buffers and caches

ARMv4 processors and software expect to be connected to a byte-addressed memory. Word and halfword accesses to the memory will ignore the alignment of the address and return the naturally-aligned value that is addressed (so a memory access will ignore address bits 0 and 1 for word access, and will ignore bit 0 for halfword accesses). ARMv4 processors must implement some method for switching between big-endian and little-endian addressing of the memory system (if CP15 is implemented, bit 7 of register 1 controls endianness). It is IMPLEMENTATION DEFINED if the endianness can be changed dynamically.

Memory that is used to hold programs and data will be marked as follows:

| | |
|---|---|
| Main (RAM) memory | will normally be set as cacheable and bufferable |
| ROM memory | will normally be set as cacheable, and will be marked as read only (so the bufferable attribute is not used, and SHOULD BE ONE) |

#### Write buffers

An ARMv4 implementation may incorporate a merging write buffer, that subsumes multiple writes to the same location into a single write to main memory. Furthermore, a write buffer may re-order writes, so that writes are issued to memory in a different order to the order in which they are issued by the processor. Thus IO locations should never be marked as bufferable, to ensure all writes are issued, and in the correct order, to the IO device.

#### Caches

Frame buffers may be cacheable, but frame buffers on writeback cache implementations must be copied back to memory after the frame buffer has been updated. Frame buffers may be bufferable, but again the write buffer must be written back to memory after the frame buffer has been updated.

ARMv4 does not support cache coherency between the ARM and other system bus masters (bus snooping is not supported). Memory data that is shared between multiple bus masters should be mapped as uncacheable to ensure that all reads access main memory, and unbufferable to ensure all writes do access main memory. IO devices that are mapped into the memory map should be marked as uncacheable and unbufferable.

The coherency of data buffers that are read or written by another bus master may be managed in software, by cleaning data from writeback caches and write buffers to memory when the processor has written to the data buffer and before the other bus reads the buffer, and flushing relevant data from caches when the buffer is being read after the other bus master has written the buffer. An uncached, unbuffered semaphore can be used to maintain synchronisation between multiple bus masters. See *7.14 Semaphores* on page 7-31.

---

# System Architecture and System Control Coprocessor

For implementations with writeback caches, all dirty cache data must be written back before any alterations are made to the MMU page tables, to ensure that cache line write back may use the page tables to form the correct physical address for the transfer.

Caches may be indexed using either virtual or physical addresses. Physical pages must only be mapped into a single virtual page, otherwise the result is unpredictable. ARMv4 does not provide coherency between multiple virtual copies of a single physical page.

Some ARM implementations support separate instruction and data caches. The coherency between the data and instruction cache may not be maintained in hardware, so if the instruction stream is written, the instruction cache and data cache must be made coherent. This may entail cleaning the data cache (storing dirty data to memory), draining the write buffer (completing all buffered writes), and flushing the instruction cache. Instruction and data memory incoherency occurs after a program has been loaded (and thus treated as data) and is about to be executed, or if self-modifying code is used or generated.

## 7.13.2 Interrupts

ARM processors implement fast and normal levels of interrupt.

Both interrupts are signalled externally, and many implementations will synchronise interrupts before an exception is raised. A fast interrupt request (FIQ) will disable subsequent normal and fast interrupts by setting the I and F bits in the CPSR, and a normal interrupt request (IRQ) will disable subsequent normal interrupts by setting the I bit in the CPSR. See *2.5 Exceptions* on page 2-6.

### Cancelling interrupts

It is the responsibility of software (the interrupt handler) to ensure that the cause of an interrupt is cancelled (no longer signalled to the processor) before interrupts are re-enabled (by clearing the I and/or F bit in the CPSR). Interrupts may be cancelled with any instruction that may make an external data bus access; that is, any load or store, a swap, or any coprocessor instruction.

Cancelling an interrupt via an instruction fetch is UNPREDICTABLE.

Cancelling an interrupt with a load multiple that restores the CPSR and re-enables interrupts is UNPREDICTABLE.

Devices that do not instantaneously cancel an interrupt (i.e. they do not cancel the interrupt before letting the access complete) should be probed by software to ensure that interrupts have been cancelled before interrupts are re-enabled. This allows a device connected to a remote IO bus to operate correctly.

**ARM Architecture Reference Manual**

ARM DDI 0100B

## 7.14   Semaphores

The Swap and Swap Byte instructions have predictable behaviour when used in two ways:

- Multi-bus master systems that use the Swap instructions to implement semaphores to control interaction between different bus masters.

  In this case, the semaphores must be placed in an uncached and unbufferable region of memory. The Swap instruction will then cause a (locked) read-write bus transaction.

  This type of semaphore may be externally aborted.

- Systems with multiple threads running on a uni-processor that use the Swap instructions to implement semaphores to control interaction of the threads.

  In this case, the semaphores may be placed in a cached and bufferable region of memory, and a (locked) read-write bus transaction may not occur.

  This system is likely to have better performance than the multi bus master system above.

  This type of semaphore has unpredictable behaviour if it is externally aborted.

Semaphores placed in non-cacheable/bufferable memory regions have UNPREDICTABLE results. Semaphores placed in cacheable/non-bufferable memory regions have UNPREDICTABLE results.

# Index

# Index

# Numerics

# A

# Index

**ARM Architecture Reference Manual**
ARM  DDI 0100B

**ARM Architecture Reference Manual**
ARM DDI 0100B

# Index

**ARM Architecture Reference Manual**
ARM  DDI 0100B

**ARM Architecture Reference Manual**
ARM DDI 0100B

# Index

**ARM Architecture Reference Manual**
ARM DDI 0100B

# Index

**ARM Architecture Reference Manual**
ARM DDI 0100B

# Index

**ARM Architecture Reference Manual**
ARM  DDI 0100B